Applications of Wet-End Paper Chemistry

Second Edition

Ian Thorn • Che On Au
Editors

Applications of Wet-End Paper Chemistry

Second Edition

Springer

Editors
Ian Thorn
Eka Chemicals Ltd.
Somerset
UK
ian.thorn@akzonobel.com

Che On Au
Eka Chemicals Ltd.
Somerset
UK
cheon.au@akzonobel.com

ISBN 978-1-4020-6037-3 e-ISBN 978-1-4020-6038-0
DOI 10.1007/978-1-4020-6038-0
Springer Dordrecht Heidelberg London New York

Library of Congress Control Number: 2009929354

Printed on acid-free paper

Springer is part of Springer Science+Business Media (www.springer.com)

In Memory of Hammy Singh

Foreword

Now more than ever, in an era of economic uncertainty, constrained resources and environmental sensitivity, paper producers are facing increasing challenges. Over the last decade, we have witnessed a growing interest in the field of wet-end chemistry.

One of the driving forces in papermaking is the customers' demand for superior paper quality. Economically, quality improvement has to be achieved without cost increase. This clearly indicates that issues like filler increase, reduction of water and energy consumption, and performance improvement have become vital to all of us.

New types of fillers, retention aids, biocides, sizing agents and many more additives offer a sound basis for process improvement. Chemical technologists have to critically select the most appropriate additives for their production processes and carefully watch the impact on quality, costs and environment.

A better understanding of wet-end chemistry has already led to significant progress. Benchmarks of products and processes demonstrate an increase both in quality and in environmental performance, which is not solely caused by better technical equipment but also by sophisticated utilisation of additives.

Sometimes the path from lab experiments to successful implementation of new chemistry concepts on paper machines is a difficult one, and interactions between the many constituents of the wet-end system have to be taken into consideration. A striking example is the accumulation of salts and colloids caused by the ever-increasing closure of water loops.

Not all trials result in immediate success but they contribute to new knowledge and even better ideas. Close collaboration between suppliers, scientists and papermakers results in better solutions and reduces the risk of failure.

This book is written by experts in their field and it provides comprehensive insight into latest research and applications. The combined expertise of the authors of this book offers new opportunities to foster wet-end efficiency and to increase product quality.

Heidemarie Reiter

Contents

Contributors

Che On Au
Global Application Manager, Eka Chemicals Ltd., Bridge House, High Street, Congresbury, Bristol, BS49 5JA, UK
Cheon.Au@akzonobel.com

Roland Berger
Product Manager Mütek Lab & Online, BTG Instruments GmbH, Arzbergerstrasse 10, D 82211 Herrsching, Germany
roland.berger@btg.com

M.T. Crisp
Ashland Hercules Water Technologies, Noordweg 9, 3336 LH Zwijndrecht, The Netherlands

Paul Cutts
Solution Application Specialist, Ciba Specialty Chemicals, Cleckheaton Road, Low Moor, Bradford BD12 OJZ, W. Yorkshire, UK

Peter de Clerck
Regional Marketing Manager Pulp and Paper Chemicals, Connell Bros. Co. Ltd., 345 California Street, 27th Floor, San Francisco, USA CA94104;
Connell Bros. Co. (Thailand) Ltd., B.B. Building, 8th floor, Unit 18–02\No.54, Sukhumvit, 21 Road (Asoke), Klongyoey-Nuea, Wattana, Bangkok 10110, Thailand
pdeclerck@connellthai.com

Thord Hassler
BIM Kemi AB, Box 3102, Hunstugans vag 7, 443 03 Stenkullen, Sweden
thord.hassler@bimkemi.com

Philip M. Hoekstra
Director Systems and Support, Buckman International
pmhoekstra@buckman.com

Danny Ingle
Technical Marketing Manager, Asia Pacific, Imerys Asia Pacific PTE Ltd., 80 Robinson rd 19-02#, Singapore 068898

Stan Jenkins
Papertex Ltd., P.O. Box 11, Skipton, BD23 9AF, UK

Dave Johnson
Eka Chemicals Ltd., Bridge House, High Street, Congresbury, Bristol BS49 5JA, UK

Leslie McLain
Technical Manager – Uncoated Applications, Imerys, Pigments for Paper, North America, Roswell, GA, North America

Barry McLow
3 Buxton Lane, Droysdon, Manchester M43 6HL
barrymclow@hotmail.com

Sal Mirza
Assistant Professor, Midway College, Midway KY, USA 40347
smirza@midway.edu

Richard J. Riehle
Ashland Hercules Water Technologies Inc., Research Center, 500 Hercules Road, Wilmington, DE 19808, USA

Ian Thorn
Business Manager, Central European Sales, Eka Chemicals Ltd., Bridge House, High Street, Congresbury, Bristol, BS49 5JA, UK
ian.thorn@akzonobel.com

Chapter 1
Introduction

Ian Thorn and Che On Au

Background

We published the first edition of this book in 1995 and in the ensuing years change in the pulp and paper industry has continued at great speed. Pulp production, for instance, is moving from the forests of the USA and Nordic countries to South America. Similarly, new capacity in papermaking is moving east to Asia from North America and Europe.

The last decade has seen much consolidation amongst both papermakers and their suppliers. In order to survive, the industry has had to rethink its strategies. Cost has become the leading topic of boardrooms throughout the world. Whether this means raw materials, energy, employee numbers or new investment, pressure on existing resources and those serving the industry becomes greater.

Hence, suppliers have had to provide products of greater efficiency to meet this challenge. Briefly, we have experienced changes in machinery (headbox technology, self-feeding dryer sections and improved clothing fabric design). The closure of water circuits have continued apace. In pulping, the use of chlorine for bleaching has almost disappeared and a far greater proportion of hardwood is used in fine paper than ever thought possible. This has been coupled with calcium carbonate filler contents of 25–30%.

Clearly then, paper chemicals must have changed to accommodate this evolution in papermaking. The authors of the following chapters are experts in their fields and will demonstrate in greater detail what has happened since 1995. Some paper grades have not changed and neither has the way they are treated chemically.

I. Thorn
Business Manager, Central European Sales, Eka Chemicals Ltd., Bridge House, High Street, Congresbury, Bristol, BS49 5JA, UK
e-mail: ian.thorn@akzonobel.com

C.O. Au
Global Application Manager, Eka Chemicals Limited, Bridge House, High Street, Congresbury, Bristol, BS49 5JA, UK
e-mail: Cheon.Au@akzonobel.com

I. Thorn and C.O. Au (eds.), *Applications of Wet-End Paper Chemistry*,
DOI 10.1007/978-1-4020-6038-0_1, © Springer Science+Business Media B.V. 2009

Others have changed radically to allow faster production and improved quality. Times have changed since the first edition and in response to this change we felt that an update was required.

Process Chemicals

Paper chemicals fall generally into two categories: process chemicals which support the production of paper (e.g. slimicides, defoamers, drainage/retention aids and others like machine clothing cleaners, which are not covered by this book) and performance (or functional) chemicals. The latter are used to improve paper properties such as dry strength, wet strength, colour, hydrophobicity and the ease with which it can be printed upon.

Deposit Control

Despite the ongoing evolution of paper machine technology no engineering solution has yet been found for the problem of deposits. Part of the problem is that paper is still made predominantly from the pulping of trees and water from a nearby source. Even after pulping and bleaching, the fibres can still have pitch particles associated with them and similarly, untreated water may contain bacteria, spores and fungi.

If these various chemicals and organisms are allowed untrammelled existence, deposits will result.

The effect of these deposits varies considerably between spores in food-packaging paper, potentially causing a health hazard, organic deposits breaking away from the surface of a pipe to form a spot in the paper or cause a paper break.

No deposit can exist without a surface but there is no shortage of surfaces in a paper machine approach flow, wet-end and dyer section. Indeed thousands of square meters exist.

Not all surfaces can be engineered to avoid deposits. Although pipes and headboxes can be polished, these too will eventually become the nexus for a layer of scale. Other areas of the machine such as a machine wire will attract a different type of deposit.

The measurement of the concentration of some of the troublesome organisms is often a precursor to the treatment of the problem.

In the case of microbiological organisms which cause slime and subsequent breakaways, microbicides are used. These have changed in the last 10–20 years in order to comply with recent legislation. Products used today, e.g. chlorine dioxide, have to break down into non-toxic chemicals after they have killed the micro-organisms.

Minimising organic deposits, though, can be achieved by the use of low molecular weight organic polymers.

The manner in which the four different types of deposit are formed and how they are treated is described in detail in Chapter 2.

Retention, Drainage and Formation

Retention, drainage and formation were one of the most difficult balances to obtain in papermaking when only one product was used to attain these properties. When a few hundred grams of polyacrylamide were used to flocculate the stock just before the headbox, both the drainage and fines retention increased as more was added. Similarly, paper formation deteriorated with increasing retention aid addition.

Retention systems have become increasingly more sophisticated as our understanding of wet-end chemistry has improved and paper machines have become more demanding. Often two, three or even four chemicals are necessary to attain the retention, drainage and formation balance on modern machines. This sometimes involves a coagulant to neutralise anionic trash, then silica sol or bentonite, anionic or cationic polyacrylamides (or even both) and cationic starch to flocculate the stock.

Often the paper grade itself determines the retention system used although this too is changing as chemical pricing is affected by the aging of patent protection. Chapter 3 explores the uses of these chemicals and how troubleshooting can be effected.

Foam

As long as paper is produced with fibres and water, foam will always potentially exist. The greater the turbulence experienced by the stock, the greater the chances of air being entrained. Unfortunately, modern trends accentuate this problem as machine speeds increase and greater use is made of gap former technology.

Some of the problems that foam and entrapped air can cause are the cavitation of pumps creating runnability problems, slowing of the drainage through the sheet, deposit formation in the press and dryer sections and the formation of pinholes in the sheet.

Chapter 4 gives a thorough explanation on how the potential for these problems can be measured, the chemicals used to solve them and several case histories illustrating troubleshooting techniques.

Performance Chemicals

Sizing

Perhaps one of the greatest changes in technology witnessed since the last edition is the decline in the use of anionic rosin size in Europe. Imparting hydrophobicity to paper by the use of rosin modified by maleic anhydride or fumaric acid and stabilised by casein at pH 4.5–5.0 is now seldom seen.

In fine paper, in particular, the change was prompted by the increased use of calcium carbonate fillers at the expense of kaolin. Far more calcium carbonate could be incorporated into the sheet than clay and now some papers contain in excess of 30% of this filler. Once calcium carbonate was substantially in use, users of waste paper could no longer size at low pH without foam and related problems.

Alternative types of size quickly filled the void, rosin was reformulated to enable sizing up to pH ~6.5, whilst alkyl ketene dimer (AKD) and alkenyl succinic anhydride (ASA) provided non-rosin solutions.

Whilst cationic rosin size (CRS), AKD and ASA have all found uses within the industry, e.g. AKD and CRS in liquid packaging board and ASA in gypsum board, perhaps ASA has gained the greater market share.

The initial problems posed by the use of this material have been largely solved. The possibility of deposits is still greater than when using AKD or even CRS in some cases and the equipment used to emulsify ASA is expensive compared to pumping ready-to-use AKD or CRS. Nevertheless, the cost of sizing is often lower.

The pros and cons of these and other materials is extensively reviewed in Chapter 5, where also the troubleshooting section is supported by nearly 40 years of experience with these products in our industry.

Fillers

Filler is used to describe minerals that augment fibre in the final sheet. In the section above it was suggested that sizing trends were at least partially caused by the choice of filler. However, kaolin and calcium carbonate are not the only choices. For example, talc, precipitated silica and titanium dioxide are also used depending on machine conditions and final characteristics demanded of the final paper.

Initially, fillers were used to replace more expensive fibre without detrimentally affecting the properties of the sheet.

Since then, with the increasing sophistication of papermaking, the role of fillers has dramatically changed. Although the commercial value of using fillers still exists, a range of optical physical properties can be imparted to the sheet.

In addition, the use of fillers also affects such parameters as chemical usage, drying and formation as well as brightness and printability.

The properties, choice and application of fillers are expertly covered in Chapter 6.

Dry Strength Resins

Dry strength resins are used to impart strength to the sheet much like cationic starch. They are used to increase the bonding area between the fibres and augment the hydrogen bonding which holds the sheet together in its dry state.

Dry strength resins have existed since the 1950s and were predominantly in use in Korea and Japan where starch was mostly regarded as a food source. In addition, low molecular weight polyacrylamides were readily available to the extent that they could be manufactured nearby and pumped to the paper machine.

Since then, the number of products available has grown and become more economically competitive with cationic starch. Since the last edition of this book new chemistry has become available to the industry. Both glyoxylated polyacrylamides and polyvinylamines are now being used. Chapter 7 provides up-to-date information on these and other products as well as case histories and troubleshooting tips.

Wet Strength Resins

Wet strength resins unlike dry strength resins are added predominantly to impart strength to paper when it becomes wet. The chemistry, development and application of the predominant products are reviewed in Chapter 8 in some detail.

These products are used on a number of grades: hygiene papers, some liquid packaging grades, wall paper base, sack grades, etc.

The greatest changes seen in wet strength resins have been prompted by European legislation, particularly when the paper grade in which it is used comes into contact with hot drinks. These products include tea bags and coffee filters. Great ingenuity has been shown by major suppliers in reducing almost to zero certain chlorine-containing by-products of polyaminoamide-epichlorohydrin resins.

Wet-End Starch

Starch use in papermaking is nearly as old as the use of fibre and water alone. Now, globally, nearly a million tonnes are added to the wet-end every year.

Papermaking trends have had their effect on cationic starch formulations too. In response to the closing up of white-water circuits (to reduce water usage), the resulting increase in specific conductivity of the water has required increasingly higher cationic charge on the starch.

Chapter 9 provides a broad review of the evolution of starch use in paper to the present. The chemistry of starch components and their contents in starch from various sources is covered in detail. Although cationic starch is predominantly the product of use, the use of both anionic and amphoteric starch (which contains both anionic and cationic groups) is also mentioned.

Although cationic starch was used primarily to improve the strength of paper, over the years other benefits have been utilised. Cationic starch has found use as a retention aid component when paired with colloidal silica. In sizing too cationic starch has been used, and both AKD and ASA have been emulsified with it.

Dyeing

One of the true arts, still practised in papermaking, is the colouration of paper. Chapter 10 covers this subject with great delicacy. Explanations are given of how colour is perceived by the human eye and how differences in shade are distinguished. The chemistry of various chromophores is discussed as well as the conditions under which these chemicals are best used: e.g. acid dye use in crêpe paper, basic dyes in newsprint and direct dyes in printing and writing papers.

The chapter goes on to show how the dyes are applied, what other chemicals can interfere with this process and how these problems can be avoided.

Online Measurement and Control

The number of chemicals used to maintain the balance on the paper machine of both process chemicals, e.g. three or four component retention aids, deposit control chemicals as well as defoamers and anionic trash catchers (ATC) and performance chemicals, e.g. cationic starch, size, filler and perhaps wet strength resin and/or dye, has resulted in a wet-end 'soup'.

Despite this, the control of wet-end chemistry has in recent years, become somewhat simpler.

The one thing that the chemicals mentioned above have in common is that when dissolved in water the surface of the dissolved chemical develops a cationic or anionic charge. For example, an ATC by the nature of its function must hold a cationic charge in order to neutralise anionic material originating from pulp or coated broke, whilst bentonite and colloidal silica both carry a negative charge so that they can interact with positively charged polyacrylamides.

Although the ability to measure the final charge of the white water or stock in the laboratory has been possible for some time this only gives a 'snapshot' of the wet-end condition. The advent of the online measurement has enabled the result of existing chemistry to be viewed on a screen continuously and the running of trials of new chemicals far easier.

Chapter 11 also reviews the online measurement of retention, free air and ash. It also explains with case histories how monitoring and control have been utilised.

The view that we held in 1995 has not changed in that we believe that the wet-end should be in the hands of the papermaker rather than the supplier. With that view we have asked authors of outstanding practical knowledge to contribute to this book, which we hope enjoys the same success as the last edition.

Chapter 2
Improving Productivity and Quality by Controlling Organic, Inorganic and Microbiological Deposits

Thord Hassler

2.1 Introduction; the Significance of Appropriate Surface Quality

Paper can today be made at a speed above 2,000 m/min. Of all global industrial manufacturing processes paper production is the most automated. The operator sits behind three-glass windows surveying the process on monitors. Around the machine there is no one, and the monster, high as a three-floor building, is humming on. Paper manufacturing is on its way to a world not far from Matrix, Star Wars or any of the other dream productions from Hollywood. It is easy to forget that paper comes from a tree, and the tree was not created to be made into paper. And there is a tendency to ignore that the world is not as perfect as a movie film where everything can be erased with digital techniques.

The modern papermaking process is very sensitive to disturbances. The formation of deposits is one of the most important factors limiting the productivity on a paper machine today. Further, paper defects (such as breaks, holes and spots very often caused by deposits) represent one of the most common complaints of a user/buyer of paper. Still paper machine deposit control as a means to improve paper quality and productivity does not get appropriate attention. On the contrary, deposit control is often considered unnecessary in the ultimate papermaking process, which all papermakers strive to reach. Academic research, with few exceptions, ignores the subject.

Needless to say, for a deposit to develop at a minimum two things are required: a surface and some material that has the potential to deposit. Before addressing the deposit it is appropriate to consider the different surfaces in the pulp and paper industry, and the significance of these.

In a pulp and paper mill there are thousands of square meters of surface. These surfaces comprise chests and pipes, wires and felts, drying cylinders and dryer

---

T. Hassler
BIM Kemi AB, Box 3102, Hunstugans vag 7, 443 03 Stenkullen, Sweden
e-mail: thord.hassler@bimkemi.com

I. Thorn and C.O. Au (eds.), *Applications of Wet-End Paper Chemistry*,
DOI 10.1007/978-1-4020-6038-0_2,

fabrics, cleaners and refiners, pH and consistency meters, press rolls, lead rolls and so on. The properties of these surfaces vary vastly. A press roll should have very different properties compared to a drying cylinder. A wire should have certain properties and headbox should have fundamentally different surface properties for optimal functionality. It is not that the ideal properties of a surface are not known. A headbox should be as smooth as possible, which is why this is electropolished at manufacture. A centre press roll should have a certain roughness.

However, not even in the design is any surface perfect. In order to come acceptably close to the ideal properties various materials ranging from basic stainless steel to space age material are used. Polyester is an appropriate material for wires, while polyethylene has been found to be an inferior construction material. Teflon is a good construction material in many aspects, but is too soft and not sufficiently durable. Foils should preferably be made of as hard a material as possible. Ceramics are used, but not always as such material is expensive. Titanium would be a good material for certain machine parts but is never used since it would be too expensive. Constructing a paper machine means making compromises between different properties of materials taking cost into consideration.

Surfaces are also worn with time. An abrasive doctor blade ruins a drying cylinder surface. Corrosion deteriorates any surface. Plastic materials wear rather quickly, but can be easily replaced. A dropped screwdriver can cause serious damage anywhere. With time the properties of surfaces just get progressively worse.

Properties of surfaces deteriorate as deposits build up. A thin invisible film of pitch can transform the original almost-perfect surface of a press roll to give much more inferior properties. The smooth surface of an electropolished headbox is transformed to a material with high roughness through the deposition of a thin layer of scale.

A lot of production limitations in the paper industry lie in limitations in surfaces. The idealist ignores these limitations, as surfaces always 'should' be perfect. There 'should' be perfect release from the centre press roll; there 'should' not be a build-up of deposits on the forming fabric. The consistency meter 'should' give the correct result, but it does not because something has deposited on a crucial place. There 'should' not be threads formed in the headbox. The idealist is frustrated as he/she cannot take advantage of the marginal production due to disturbances caused by inadequate surfaces. He/she simply forgets that fibres are living material, and not created to be made into paper.

This, you may think, is only of theoretical significance and possibly relevant comments for small machines in countries that are not considered to be 'at the forefront of technology'. Take a look at the two photos below taken of the headbox of a state-of-the-art paper machine in the Nordic countries before and after a cleaning (Fig. 2.1). Remember that a headbox should ideally be of high smoothness.

This chapter reviews the source and control of paper machine deposits and the alternative methods of application of chemicals to minimise/eliminate these troublesome deposits. It is intended to give pragmatic guidelines for the realist working to maximise productivity and quality in real-life situations.

Fig. 2.1 Pictures before and after headbox cleaning

2.2 Types of Deposits

Deposits can be broadly categorised into four different groups:

1. Microbiological deposits
2. Organic deposits of water-insoluble components
3. Inorganic deposits
4. Precipitation of organic water-soluble components

These four different types of deposits require different approaches. In this chapter they will be discussed individually. However, in reality often components of more than one category are found in a deposit, and the optimal approach is usually a combination of treatment methods. Further, one material can deposit in very different positions requiring different remedial strategies. To make it even more complicated a deposit of a certain material in one position can be largely ignored while another deposit of the same material in a different position may cause serious problems. It may sound as though the issues of deposit control are too complicated to address. It is complicated, but analysing the situation correctly, understanding the mechanisms of deposition and knowing the various methods available enable a person to make improvements in productivity and quality by controlling these deposits.

2.2.1 Microbiological Deposits

A papermaking system represents an ideal environment for microbiological growth. There is a continuous supply of nutrients. Temperature, pH, salt concentration, etc. are all well within the range for microbiological growth. The microbiological activity may result in a number of different problems for the paper industry. The following represent some examples of these problems:

1. Deposits may be formed in the system, and these deposits may then detach to either form a defect in the paper or cause a break on the paper machine.
2. Fungi and certain bacteria can produce spores that will survive the drying process, and these spores can be present in the final product, such as a packaging board, representing a health risk.
3. Certain bacteria (particularly anaerobic sulphate-reducing bacteria) can cause corrosion of the papermaking equipment.
4. Micro-organisms can form obnoxious substances through their metabolism, and these substances may create a bad odour within the paper mill itself, in the effluent and in the paper. An example of such a metabolite is butyric acid.
5. Micro-organisms may degrade the quality of papermaking raw materials, such as the darkening of filler slurries and the decrease in viscosity of starch solutions.
6. Explosive gases (e.g. hydrogen sulphide and methane) may be formed by anaerobic microbiological growth.

Various means are available to the industry to minimise all of these problems. This chapter deals only with microbiological deposits and methods to control these.

Any surface subjected to water-containing micro-organisms will sooner or later become coated with a layer of bacteria. Deposits that contain a high amount of micro-organisms are referred to as (biological) slime or biofilm, because of their slimy or gel-like character. Slime formation is most obvious in areas where the water flow is insufficient to dislodge the accumulation. The process of biofilm formation is known as biofouling. Biofilms occur in most water systems, and the pulp and paper industry is just one of many industries where this occurs.

2.2.1.1 Micro-organisms

Biofilms become after some time complex ecosystems where various micro-organisms live together in a symbiotic state or in a predatory state. The lowest form of micro-organisms can be divided into three distinct groups, called bacteria, fungi and algae. Fungi can then further be divided into yeast and mould. Most problems encountered in the papermaking process are initially associated with bacterial activity. Fungi may be present in deposits while algae are never present in a papermaking system since these require direct sunlight to grow. In mature deposits higher organisms, such as nematodes, may be present.

Bacteria are single cell organisms found just about everywhere on earth, e.g. in soil, air, water and in the digestive systems of all creatures. They are measured in microns 0/1,000 mm, and can be examined under a microscope at a magnification of 500–1,000 times. They have different shapes and colour and may grow in different forms. Microphotographs of some common bacteria found in pulp and paper systems are shown below (Fig. 2.2).

The individual species can be determined by a skilled microbiologist. However, this process is tedious, and it is generally not useful to determine the micro-organisms beyond identifying whether they are bacteria or fungi, and this information is not

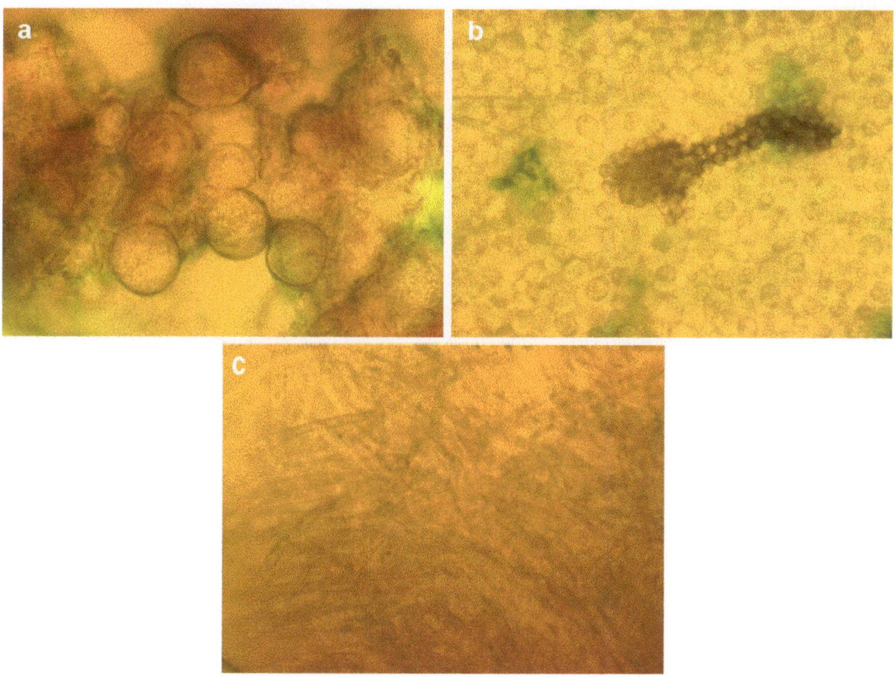

Fig. 2.2 Three microphotographs of bacteria

useful in the treatment programme. For the sake of completeness of this chapter on deposit control, some of the most common micro-organism families associated with microbiological deposits are listed below:

Bacteria

Acinetobacter
Aeromonas
Alcaligenes
Bacillus
Staphylococcus
Serratia
Klebsiella
Micrococcus
Pseudomonas
Desulfotomaculum
Beggiatoa
Gallionella
Sphaerotilus
Desulfovibrio
Enterobacter

Fungi

Aspergillus
Ciadosporium
Fusarium
Geotrichum
Penicillium
Phialophora
Phoma
Pullularia
Trichoderma
Mucor
Rhizopus

Yeast

Rhodotorula
Geotrichum
Trichosporon
Cryptococcus
Candida
Growth of Micro-organisms

Bacteria reproduce by cell division. One cell becomes two, two becomes four, and four becomes eight. Bacterial growth refers to an increase in number and not in size. The time between each cell division varies between species but is typically 10–20 min given the optimal growth conditions. At a generation time of 10 min, 1,000 bacteria would become 4 million in 2 h. It is evident micro-organisms can have a very rapid growth indeed.

Most fungi are common air and water contaminants which under favourable growth conditions can form large fibrous masses. Fungi exist as parasites and can tolerate a wide range of both temperature and pH. Fungi can be subdivided into moulds and yeasts. Moulds reproduce by releasing spores from their filaments, and grow by elongation of threads of hyphae, which produce a complex branched mass called a mycelium. Hyphae are much larger than bacterial cells, with an average diameter of 5 μ and several hyphae in the mycelium terminate in spore heads. Hyphae penetrate into the nutrient sources to obtain food and are capable of assimilating an extraordinarily wide range of organic materials. They are particularly adept at utilising complex carbon compounds and are frequently found growing on cell debris left after bacterial degradation. Yeasts are non-filamentous fungi, they are unicellular and multiply by budding, and less commonly found in deposits in pulp and paper deposits.

Growth of micro-organisms, like that of plants or animals, cannot take place unless water is present to transport nutrients and remove waste products. Water is also necessary to facilitate the many chemical reactions which take place within the micro-organisms. Given the presence of water, there are five major conditions which affect the growth of micro-organisms:

- Moisture
- pH
- Temperature
- Nutrients
- Oxygen level

All micro-organisms have an optimal pH for growth with most bacteria preferring a pH around 6.0–8.0, but they can grow at pH 4.0–10.0. Fungi prefer a slightly acid environment, with pH 5.5 being optimal, but pH 2.0–8.0 can support growth.

Micro-organisms have a temperature range within which growth will occur. There is a minimum temperature below which growth cannot take place and likewise a maximum above which growth is not possible. However, a lack of growth does not indicate the death of the organism. Somewhere between the maximum and minimum will be an optimum at which vigorous growth occurs and bacteria can be classified roughly according to the temperature at which they may grow.

Fungi can grow over a wide temperature range (0–45°C), but optimum growth normally occurs between 20°C and 30°C.

Nutrient substances and a number of trace metals are required for microbiological growth. The most important energy sources for bacteria in the paper industry are polysaccharides and their degradation products.

Fungi exist as parasites and derive their food from living plants or organisms, or can exist as so-called saprophytes nourishing from non-living organic sources.

Most bacteria require oxygen to grow and are classified as aerobes, but some flourish in the absence of oxygen and are classified as anaerobes. Bacteria which can grow either with or without oxygen are known as facultative anaerobes. Fungi are aerobic or facultative anaerobic micro-organisms.

2.2.1.2 The Five Stages of Biofilm Formation

The formation of biofilm in any system or on any surface is a very dynamic process which involves a sequence of distinctly different events. The stages of biofilm formation are depicted in the following schematic presentation (Fig. 2.3).

Stage One: Formation of the Base Layer for Biofilm Growth

On a completely clean surface, such as one following an effective cleaning operation when all deposits have been removed, the first step in the formation of a biofilm actually does not involve any microbiological material. Rather this is the establishment of a suitable layer for micro-oganisms to grown on. This layer is in the paper industry generally a layer of wood extractives or solvent extractable material entering the process through the recycled fibre. This conditioning layer can be considered making surfaces more favourable for micro-organisms to colonise.

Stage Two: Colonialisation of the Surface

When a suitable base layer has been created on the surface, micro-organisms attach to the surface. These may be selection of bacteria present in the process water which preferably grow on surfaces. This is a reversible stage; bacteria may stay or may leave.

Stage Three: Biofilm Formation

Micro-organisms which remain on the surface will become irreversible attached to it. Once the micro-organism has attached itself to the surface it will start to grow, and the rate of this growth will depend on the conditions for growth present.

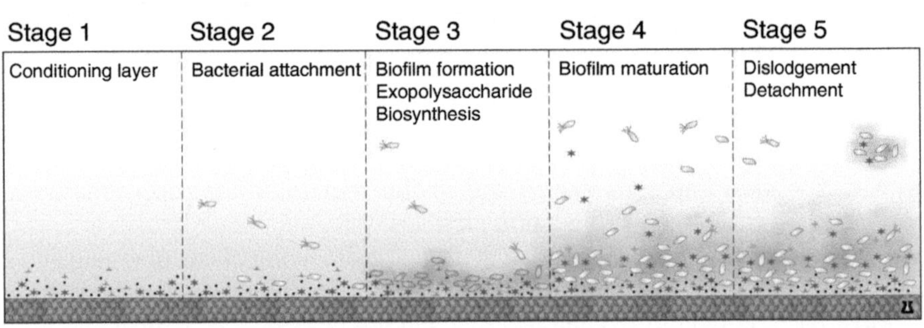

Fig. 2.3 Stages of biofilm formation

Coherent slime is formed if the micro-organism produces extracellular polysaccharides (EPS), i.e. polysaccharides produced outside the cell wall. A thick resistant matrix, which is the frame of the biofilm, is formed in this process. The chemical composition of this polysaccharide is very specific and varies from species to species, and may be a pure polysaccharide comprising a single sugar, such as glucone or levane. However, more commonly it is built up of several different sugar units. Sometimes amino sugars and amino acids are also built into the EPS. This polysaccharide can bind large quantities of water, producing a highly hydrated layer around the micro-organism which protects the micro-organism against dehydration as well as attack of higher organisms, such as amoebae and bacteriophages. This extensive hydration gives deposits of microbiological material their characteristic slimy appearance. The water content of such a microbiological deposit usually exceeds 90%.

It is often common to find also other components (like fibres and fillers) in a microbiological deposit. Further, the extractive content of a microbiological deposit can be significantly higher than the extractive content of the pulp or the process water showing an accumulation of these extractives in the total deposit. In fact, the fibre and fines part of this type of deposit have been physically incorporated in the mass where the microbiological slime acts as a binder of these components. This extractive part of the deposit may well contribute in itself to the development and growth of the deposit as well as acting as a binder and as nutrient to the micro-organisms.

Microbiological deposits may be formed in many different places in the papermaking process. However, the closer to the wet-end the more detrimental the deposits will be for the papermaking process. Microbiological deposits may form on submerged surfaces as well as surfaces above the water line of the paper machine. In the latter case it is a requirement for slime formation that a surface is kept moist by splashes or aerosol formation. Examples of such surfaces are the frame of the wire section and the frame of the press section.

Stage Four: Biofilm Maturation

As the microbiological slime ages it also changes in composition. Conditions in the interior of the slime may become anaerobic as a result of slow diffusion of oxygen within the slime deposit and anaerobic bacteria may take over within the slime, particularly close to the surface onto which the slime is formed as this is the point of lowest oxygen concentration. At this stage the microbiological deposit may cause extensive corrosion of the papermaking equipment and may produce obnoxious organic gases.

More sophisticated organisms (such as protozoa which feed on bacteria) begin to establish themselves in the biofilm as the biofilm ages and eventually nematode worms establish themselves, feeding on both organic debris and also protozoa. Using microscopic examination of slime it is often possible to establish the age of the biofilm by observing these higher organisms.

Stage Five: Dislodging of Biofilm into the Process Flow

At this point single cells to large lumps including the EPS matrix and associated particles are dislodged. The thicker the biofilm the greater the probability large lumps will be dislodged from the biofilm. Spots, holes and breaks in the paper production are all a consequence of this action. From this point on the thickness of the biofilm is generally constant.

The most difficult type of microbiological deposit is in the form which is reinforced in some way, as this is difficult to re-disperse. The most common form of microbiological reinforcement is the presence of either filamentous (thread-forming) bacteria or fungi growing in the microbiological deposit. However, reinforced or not, eventually when a biofilm becomes too thick sections will detach and contaminate the rest of the system and cause defects in the paper sheet or cause breaks.

2.2.1.3 Measuring Microbiological Activity

A number of techniques are used to follow and control microbiological activity in a paper system. These can be classified into three different methods:

- Physical
- Analytical
- Biological

Physical Methods

Physical methods can be defined as those by which the biological activity of the process water is determined indirectly. Examples of such methods used in the industry are slime board, i.e. a well-defined surface where a microbiological deposit is allowed to develop. The slime board provides a direct measurement of the current formation of microbiological deposits. While being a direct measurement it has the disadvantage that it provides a historical figure, i.e. what has happened between the last reading taken up to the time of the current reading.

A hole detector will not specifically measure biological activity, but to the extent it relates to microbiological deposits it will measure the release of biofilm and not biofilm formation.

Analytical Methods

Analytical methods can be defined as techniques to determine microbiological activity measured as a defined chemical reaction. Examples of such methods are the Resazurin test, the Ninhydrin test and the measurement of dehydrogenase activity.

Virtually all microbiological reaction pathways change the redox potential, which can also be measured. Of those listed below, there is no method that is inherently superior to any other method better correlating to machine productivity and paper quality.

A suitable redox indicator is Resazurin. The method is based on visual determination of colour change of the redox indicator with time relative to microbiological activity.

Ninhydrin is a reagent that will show the presence of amino acids, peptides and proteins through a change in colour. The method is so sensitive that even a fingerprint provides sufficient amino acids to show a positive reaction.

Dehydrogenase is an enzyme present in all living microbiological material. The presence of this enzyme can be determined using 2,3,5-triphenyltetrazolium chloride. This colourless water-soluble compound is reduced to a red compound called formasan in the presence of dihydrogenase enzyme. This compound can be extracted in ethanol, the content determined photometrically, and a relationship with microbiological activity can then be established.

Living micro-organisms produce heat, which is formed by the various chemical reactions taking place within living organisms. This heat generation can be determined by microcalorimetric methods directly related to the activity of the micro-organisms.

All living organisms use adenosine triphosphate (ATP) as a source of energy, and the energy stored in this substance can be converted into light in a complex series of chemical reactions using an enzyme called luciferase. The light emittance can be measured and the intensity of light can then be related to ATP concentration, which is related to the biological activity of the sample being analysed.

Oxygen concentration is determined electrochemically using a membrane oxygen electrode (Clark sensor). Oxygen is consumed by micro-organisms, and the reduction in oxygen consumption with time is a measure of biological activity.

Biological Methods

Biological methods can be defined as those by which the number of micro-organisms can be determined directly using cultivation methods. An example of such methods would be the determination of total bacteria count using agar broth. Biological methods are based on growth of individual micro-organisms into colonies. The number of colonies following growth in selected medium is then proportional to the original content of micro-organisms in the sample. Various media are available for selective growth of micro-organisms.

A combination of monitoring methods is often required to provide sufficient information to monitor and control the microbiological status of a paper machine. The above methods are also suitably combined with, for example, microscopic examination and micro-organism identification techniques. However, in reality, mostly physical methods are being used these days when each mill has limited resources.

2.2.1.4 Methods to Control Biofilm Formation

Microbiological growth control technologies have been used to control the problems associated with microbiological growth since the paper industry started to recycle water. Any approach to reduce the formation of biofilm will have to exercise an effect in at least one of these five stages of biofilm formation. The most common method to combat the problems caused by excessive growth of micro-organisms is the use of microbicides. In the beginning of the last century mercury compounds were used. These were replaced by biodegradable organic compounds when environmental concerns and concern for workers' safety increased. These would delay the second stage of biofilm formation, as fewer micro-organisms in the system resulted in a slower colonialisation. Using an effective biocidal product the attached organisms also grow at a slower rate, with a result that also stage three is delayed.

Several alternative approaches exist to reduce the extent of microbiological deposition on a paper machine. As with most other deposits, a combination of approaches usually produces the most cost-effective result. It is technically possible to completely prevent formation of microbiological deposits. However, this would for most systems be associated with a prohibitive cost. Instead, a microbiological deposit control programme should best be designed to control formation of deposits for a determined period of time, depending on how long the mill has to be run between stops for cleaning. This period of time could be anything between one week and two months. Three days running between regular removals of microbiological deposits does generally not require any programme to control deposition, while running for two months between cleaning requires a detailed programme, sometimes also associated with considerable costs.

Minimising System Contamination

Various methods to minimise the contamination of the paper machine system can be used. Limiting the microbiological content of water and additives going into the system is an effective method to minimise the rate of microbiological deposit formation.

Water is one of the main sources of contamination. Various oxidising compounds are used in the paper industry to treat raw water. Generally speaking, the result of the treatment is better the higher the residual content of the oxidising compound and the stronger the oxidising power of the material used. Different oxidisers are used for this purpose such as chlorine, chlorine dioxide, hypobromous acid, hypochlorite and chlorinating substances such as BCDMH. Fatty quaternary amines are used, but less often for this application because of the considerable volumes of water requiring treatment.

Papermaking additives are another potential source of contamination of the papermaking system. Such additives (e.g. filler slurries, sizing agent, retention agent and starch solutions) may contain considerable quantities of micro-organisms and, even when these additives are treated when manufactured by the addition of a preservative, they need preservation at the paper mill. This will not only minimise contamination of micro-organisms into the papermaking system, but will also maintain the additive quality, since micro-organisms may rapidly deteriorate the additive itself.

Finally, the pulp used in the papermaking process may be treated to reduce microbiological treatment requirements. It is good practice to add a preservative to the furnish during longer machine stops such as repairs, to prevent it from rotting.

Using Microbicides

Addition of microbicides to the wet-end of a paper machine is commonly used to control the growth of microbiological deposits. This should be seen as a complement rather than as an alternative to methods to reduce contamination of incoming raw materials defined above. A number of chemical substances can be used for this purpose, although the list of available substances has been reduced considerably over the last two decades as a result of more stringent approval and registration requirements. Substances used today are designed to be inherently unstable so that they break down into non-toxic degradation products for efficient disposal.

The list will be decreased as a result of the implementation of new directives, such as the European Biocide Directive that will be fully implemented by 2010. The use of biocides in paper production is often further limited by national health authorities, e.g. FDA (USA) and BfR (Europe) for food packaging paper, as well as green label lists such as White Swan. A list of substances used to control microbiological growth on a paper machine is given below:

Glutaraldehyde
Methylene bisthiocyanate (MBTC)
Potassium dimethyldithiocarbamate
Potassium N-hydroxymethyl-N-methyldithiocarbamate
Sodium dimethyldithiocarbamate
Bis(dithiocarbamate)
5-Chloro-2-methyl-4-isothiazolin-3-one and 3-methyl-4-isothiazolin-3-one
1,3-Dibromo-5,5-dimethylhydantoin
Dodecylguanidine hydrochloride
l-Bromo nitrostyrene bis(trichloromethyl) sulphone
N-alkyl dimethylbenzyl ammonium chloride
Benzothiazole-2-thiol
2-Bromo-2-nitropropane-1,2-diol 2-bromo-4-hydroxy-acetophenone
1,2-Dibromo-2,4-dicyanobutane 2,2-dibromo-3-nitrilopropionamide
3.5-Dimethyl tetrahydro-2H-1,3,5-thiadiazine-2-thione disodium ethylene)
(1,5-Pentanedial) 2-hydroxypropyl
Methanethiolsulphonate
N-4-dihydroxy-tx-oxobenzene ethanimidoyl chloride
5-Oxo-3,4-dichloro-1,2-dithiol polyoxyethylene (dimethyliminio) ethylene dichloride
Sodium 2-mercaptobenzothiazole
2-(Thiocyanomethylthio)

Typically biocides are used intermittently as this has been found to constitute a more cost-effective use of biocides. However, the optimal addition cycle and point of addition will depend on the type of biocide and the design of the system. Typically the microbicide is added 4–8 times per day for 20 min to 1 h. It is often advantageous to combine two microbicides simultaneously or alternately.

Biocides are screened and monitored using any of the test methods previously discussed. Consideration in the selection of the most suitable product should be given to the speed of action to reduce microbiological activity, the concentrations needed to obtain a reduction in microbiological activity and the persistency of the microbicide. However, laboratory screening of microbicides represents a significant simplification of the use of biocides in paper machine systems. Laboratory screening of microbicides will not provide all answers concerning which product to use and where. The experience of the supplier in the application of microbicides in paper mill systems is important in the establishment of a cost-effective treatment programme, and this will often involve trying more than one possible alternative treatment programme on the machine to then determine which the best method of application is.

Oxidising Compounds

Oxidising compounds, such as hypochlorite, hydrogen peroxide and peracetic acid, are increasingly being used, although it is relatively rare that these materials alone can control the microbiological growth. Combining an oxidising material with certain biocidal compounds has been shown to give good control. Hypochlorite used together with BCDMH is an example of an efficient two-component system.

Active oxidising compounds can also be made in a small reactor by mixing an oxidising compound with another material to produce a new compound, which is then immediately fed to the paper machine (so-called in situ generation). One example of this is ammonium bromide reacted with hypochlorite to form a substance defined as bromo-activated chloramines. A further example is the corresponding process starting with ammonium sulphate and hypochlorite as the two reactants.

Biodispersants and Biofilm Inhibitors

A number of alternative methods to control the formation of microbiological deposits not using a biocide are being increasingly used. So-called biodispersants are used in many different industrial applications to control microbiological deposits. These substances are used in conjunction with microbicides to increase their efficiency. Biodispersants perform a function by removing biofilm and making cell walls of micro-oganisms more permeable to microbicides. Substances used for this application comprise non-foaming surfactants and anionic or non-ionic low molecular weight polymers used either continuously or applied directly in relation to the addition of the microbicide.

Other products, known as biofilm inhibitors, are used as an alternative to microbicides. These products are applied without any complementary use of microbicides. For mills where environmental profile is important, this is an efficient alternative. Biofilm inhibitors have a more complex effect providing an effect in more than one of the five stages of biofilm formation. One approach makes use of three different mechanisms for biofilm control. Chemically this method of control comprises the combination of a surfactant component and a non-biocidal sulphonated lignin derivate into one formulated product. The product is applied continuously at low dosage. The surfactant component prevents, or minimises the rate of, the formation of the base layer of extractable material onto which micro-oganisms is attached. This is achieved without any detrimental effect on paper quality, foaming, retention, sizing or any other paper quality aspects or production parameters in a number of documented cases. The sulphonated lignin derivate component prevents the attachment of micro-oganisms onto the surface and interferes with the formation of the sticky EPS. Also this effect is achieved without any detrimental effect on paper quality, foaming, retention, sizing or any other paper quality aspects or production parameters in a number of documented cases. This mechanism is what is sometimes referred to as 'biodispersion'. Biodispersants work only by one mechanism while biofilm inhibition works by at least one additional mechanism.

Also such materials, even though these are typically of non-toxic nature, must be registered both in Europe (European biocide directive) and the USA, limiting the list of products commercially available.

2.2.2 *Organic Deposits*

Organic deposits can either originate from the tree itself or from man-made substances getting into the papermaking process from recycled broke or waste paper or through additives. An organic deposit originating from the tree itself is often called 'pitch', or sometimes 'troublesome pitch' to make the distinction with the fraction of pitch that does not form a deposit. Deposits of man-made origin are called either 'stickies', from their physical characteristics, or 'white pitch', from their whitish colour of styrene butadiene rubber (SBR) latex.

Natural Wood Extractives

Natural wood extractives represent chemically a very large group of components that have not been removed in the process of preparing the pulp. The potentially depositable part of these components, usually collectively called wood resin or pitch, is characterised by their lipophilic (Latin for 'oil-loving') nature being components insoluble in water but soluble in an organic solvent. There is no clear definition of this class of compounds. One definition is material that is extractable in an organic solvent with a dielectric constant of less than 3.0. This definition distinguishes wood resin from other wood extractives such as lignins, phenols and carbohydrates.

There are also more or less water-soluble hydrophilic (Latin for 'water-loving') non-fibrous components released in the manufacture of the pulp. Such components, which are not always removed in the pulping process, may disturb the papermaking, e.g. interact with additives such as retention agents. However, these water-soluble components do not form deposits other than under specific circumstances.

The content of lipophilic wood extractives (pitch) of the tree itself varies in the range of 5.0–1.0% (of dry wood), depending on the wood species and solvent used. The content of these potentially depositable substances is reduced in several stages of the pulp-making process. Ageing as well as bleaching oxidises the components, which increase their water solubility. Washing processes remove pitch from the fibres. Complete removal is however neither technically possible nor commercially justifiable. For a fully bleached chemical fibre the content of these extractives is reduced to the range of 0.2–0.1% and even below, but for unbleached grades and for mechanical fibres the content is considerably higher approaching the original content of extractives in the wood.

Chemically this group of wood component, insoluble in water but soluble in organic solvents, comprises a large group of different components. These can be divided into four groups: aliphatic components and terpenoids. Aliphatic components comprise predominantly fat, being predominantly fatty acid esters of glycerol, and their corresponding free fatty acids, but also waxes. The fatty acids are long chain aliphatic monocarboxylic acids predominantly with a chain length between 16 and 24 carbon atoms. These can be saturated or unsaturated. Linolic acid (with two double bonds) is usually dominating. In addition the aliphatic fraction may contain fatty alcohols and waxes and fatty acid esters of other alcohols and glycerols. The terpenoids also represent a comprehensive group of different chemical components containing rosin acids, e.g. abietic acid, and sterols, e.g. betulinol and beta-sitosterol.

All these lipophilic components perform a function for the living tree. The exact function is not understood for all substances. The composition of the lipophilic wood extractives varies from species to species, and is far from homogeneously distributed in the tree. All woods contain sterols, waxes, glycerides and other aliphatic extractives. However, softwood contains considerable quantities of rosin acids (e.g. abietic acid), but these acids do not exist in hardwood (e.g. birch and aspen). Certain components are predominantly located in the bark of the tree. The sterol fraction from birch and aspen bark comprises almost exclusively beta-sitosterol by way of example.

Many of the lipophilic components may also react with bleaching chemicals and are consequently modified in the bleaching process to form an even more complex mixture of wood resins.

Stickies/Inks/White Pitch

Stickies is the collective name of lipophobic components applied in the converting of paper to make the final consumer product (e.g. telephone directories, envelopes and corrugated board). SBR, vinyl acrylates, polyisoprene and polybutadiene are the most common components among the chemicals used as contact adhesives. Hot melts are always mixtures of various components, e.g. wax and tackifying resins.

The composition of ink will depend very much on the printing technique used. Adsorbing inks (e.g. used in most newsprint printing) contain typically mineral oil, unsaturated fatty acids and alkyd resins, whilst radiation curing inks (UV or IR) would be typically epoxy acrylates, urethane acrylates or similar prepolymers that can be polymerised further on irradiation. Finally xerographic inks contain yet other binders, e.g. styrene butadiene rubber or vinyl acrylates

Starch and various lattices are used as coating binders. Upon recycling (as coated broke in the mill itself or with the waste paper) the starch will not deposit while certain lattices have been found prone to deposition. Suitable lattices are SBR and polyvinyl acetate.

Additives
Deposits may also be formed from additives used in the process, e.g. defoamers, sizing agents and wax dispersants. Typical components that can be found in deposits originating from this source are: fatty alcohols (defoamer), ethylene bisstearamide (EBS) wax (defoamer), calcium salt of alkenyl succinic acid (ASA sizing agent) and paraffin wax (from wax dispersions). Other additives although by themselves not depositable, such as talc, may often be found incorporated into a deposit with mixed composition.

2.2.2.1 Analysis of Organic Deposits

Complete chemical analyses of all components of an organic deposit on a paper machine using unbleached fibres are a considerable task, and the corresponding analysis when bleached fibres are used is a virtual impossibility. When waste paper is used as fibre raw material still other organic depositable substances are introduced into the system. This also applies when coated paper is recycled as broke. A complete analysis of an organic deposit is not usually meaningful because of the complexity of components in the deposit. However, it is often worthwhile to determine the most common components of a deposit, as this may reveal the source of the material.

The first step in any analysis of an organic deposit is an extraction of the deposit with an organic solvent. Various solvents can be used such as ether, dichloromethane, chloroform or tetrahydrofuran. The composition and the content of the extractives will vary depending on the solvent used, as well as how the sample has been prepared.

The extractive of the deposit can be subjected to various analytical techniques. Due to the complexity of the composition of deposits, no singular analytical technique alone can satisfactorily cover all of the main components. Infrared (IR) analysis, or the more sophisticated FTIR, will provide information on the overall nature of the sample. Comparison of the FTIR spectra with a library of spectra is assistance in the interpretation to assign a chemical structure to the spectra.

Commercial spectra libraries are less useful, and a library built up entirely on additives and components used in the paper industry is preferred.

Often direct analysis of the sample provides limited information, as a deposit usually contains a multitude of chemical components. For a more detailed analysis it is necessary to separate the components from each other before each component/fraction is analysed and successive extraction with different solvents (e.g. first ethanol, then ethyl acetate and finally trichloro ethane) will provide additional information over extraction with one solvent only. Other fractionation techniques are available.

All the different techniques used in chemical analyses of organic material can and are being used in the analysis of deposits. Gas chromatography (GC) methods have been developed for the analysis of volatile extractive components and for analysis of non-volatile polymeric components pyrolysis GC can also be used. Size exclusion chromatography (SEC) is a convenient way of fractioning the sample, after which each fraction may be analysed with standard organic analytical techniques, e.g. nuclear magnetic resonance (NMR), C^{13} NMR and IR (of FTIR).

2.2.2.2 Physical Form of Organic Depositable Material

Organic water-insoluble depositable materials may be present in a papermaking system in five different forms:

1. Natural wood resins may exist within the fibre itself.
2. Potentially depositable organic substances may be adsorbed onto the surface of the fibre.
3. The potentially depositable material may exist in the form of colloidal particles. The size of these particles is, for natural wood pitch, slightly below 1 μ with a distribution up to a few microns. For depositable material of synthetic origin particle size may be considerably bigger.
4. Colloidal particles could be aggregated into an agglomerate.
5. The organic material may be present as a deposit in the system.

The deposition of organic material will depend very much on the form of the depositable material.

2.2.2.3 Mechanisms of Deposition

Organic deposits are formed in several different ways. Before discussing the various means of removing or negating the problems, it is necessary to review how deposits are formed and what determines the rate of deposit formation.

1. Creaming and flotation

Because all organic materials are lighter than water they may cream to the surface. The particles may coalesce into a film which is later deposited onto the wall at the air/water interface, as the concentration of particles at the air/water interface will be very high at this point of the system. Entrained air, which is often found in paper machine systems, will also enhance the accumulation of these colloidal particles onto the surface in a similar way to the flotation de-inking process. Once the entrained air bubble onto which colloidal material has been adsorbed reaches the surface it may collapse, and the accumulated particles may coalesce and form a deposit on the wall at the air/water interface.

2. Transfer from fibre to a surface
Resin adsorbed onto a fibre may be transferred from the fibre to a surface. At the same time a deposit may be removed by another fibre, and whether a deposit is formed or removed is a result of the dynamic balance that will depend on the content of depositable material and the ability of the fibre to remove/pick up deposited material. This kind of balance exists in the furnish as well as the formed sheet in contact with a press roll or a drying roll.

3. Shear deposition
Colloidal material may deposit as a result of hydrodynamic shear. There are many positions where shear is high close to a surface such as at the forming wire. Because pitch/stickies have plasticity the deposit will flow from a point of high shear to one of lower shear and, having got to this point of lower shear, there is then no mechanism by which the pitch may be removed. At this point the deposit will gradually build up.

4. Transfer from one surface to another surface
A deposit formed on one surface may be transferred to another surface at contact. Transfer of a deposit on a forming wire to foils and transfer of a deposit in a felt to an Uhle box are two examples of this deposition mechanism.

5. Deposits formed by evaporation
Deposits may form as a result of evaporation of water. Splashes of backwater onto the frame of the machine in the forming section may dry to leave a residue.

2.2.2.4 Methods to Control Deposition

Only a small fraction of potentially depositable material will form deposits on a paper machine. Most of the material that could deposit will pass through the system either ending up in the paper or in the effluent without causing any problems whatsoever. Hence, it is also of interest to understand not only how organic deposits are formed, but also the parameters determining the extent, or rather the rate, of deposit formation. Deposition is best seen as a statistical process whereby a certain, small percentage of potentially depositable substances will eventually deposit, with a majority of the material going through the process without causing any problem of deposition.

Numerous parameters affect the extent of deposition. These parameters include factors like temperature, pH, closeness of the system, fibre source, age of wood, nature of the surface the deposit is formed on, etc. It is, however, convenient to consider deposition of organic material as a function of four different parameters. Everything that can influence the extent of deposition, like those examples given above, can be covered by these four different parameters, which are:

1. *Content* of depositable organic material
2. *Depositability* of this material
3. *Colloidal stability* of the depositable material
4. *Surface affinity* of the depositable material

Deposition of organic material will be extensive if all these parameters are unfavourable. Deposition of organic material can be reduced by minimising the contribution of each of these. It is possible to address only one of these four parameters and achieve certain results. However, the best overall result is achieved by addressing two or three or all of these parameters at the same time.

Minimising the Content of Organic Material

The content of organic material is a rather obvious parameter. If there is no depositable organic material there will be no deposit, and if the content of the material is high then deposition can be expected to be significant. The content of colloidal organic material is of course the result of many different factors, the most important ones being the fibre source, the treatment of the fibres and the extent of accumulation as a result of water recycling. By way of example, chemical fibres will contain considerably less colloidal material in comparison to mechanical fibres and waste paper.

Even if it is conceptually easy to understand the significance of this parameter, it is not easy to determine what the concentration really is. Organic material may be present in a system in different forms (see Fig. 2.4 below). Obviously the organic material present inside the fibres will not make a contribution to the deposition, as it is not accessible for deposition unless it is transferred to the exterior of the fibre. In a system where exclusively fully bleached chemical fibre is used most of the

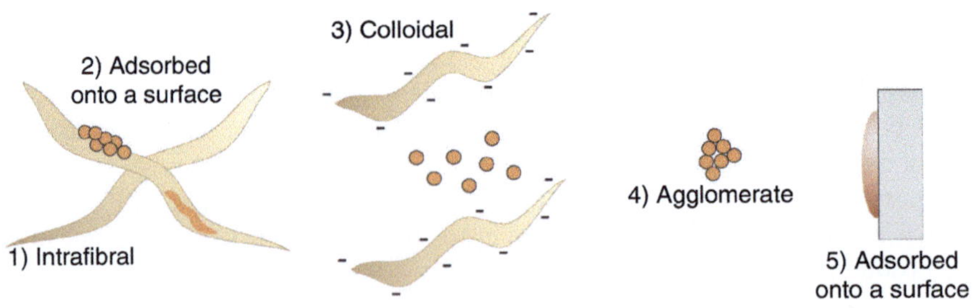

Fig. 2.4 Deposit formation

organic extractives will be in this form, and it will be erroneous to determine the content parameter as the total extractive content such as from dichloromethane.

There is significant evidence that it is the colloidal form of organic material that is the main contributor of deposits in the wet-end section of a paper machine. Methods have been developed to quantify the content of organic colloidal material in the furnish. One convenient method is to use a hemacytometer cell designed to count red blood corpuscles using a microscope. Also more sophisticated online methods have been developed typically based on image analysis techniques.

Whilst the above method is appropriate to quantify the content parameter relevant for deposition in the wet-end section, it is not relevant to use this method to quantify the content parameter for deposition in the press section, such as press rolls. Deposition on a press roll can be seen to be in equilibrium with the organic material adsorbed onto the fibres of the paper. It is clear that if the content of organic material attached onto fibres is increasing there will be an increased tendency for deposition. If a mill starts a deposit control programme using a pitch/stickies fixing agent in the wet-end, the content of colloidal organic material in the furnish will decrease as a result of the treatment, reducing the tendency for deposit formation in the wet-end of the paper machine. However, the content of organic material adsorbed onto the fibres will increase, with the result that deposition in the press section will increase if the product is not at the same time able to fix the particles more strongly to the fibres. Unfortunately there is no method available to quantify the content parameter relevant to this mode of deposition.

Ageing of Wood

Ageing as well as bleaching oxidises the components, which increase their water solubility. The material will have a higher tendency to stay in the aqueous phase. Volatile extractives evaporate further reducing the extractive content. This is what smells from freshly cut trees. Ageing of wood and chip before cooking considerably reduce the pitch problems.

Removal by Washing

Any washing operation will reduce the content of downstream colloidal material. Screening is an efficient method of removing organic contaminants from the fibres. The removal can be aided by dispersants that will shift the equilibrium of depositable material adsorbed onto the fibre to depositable material in the aqueous phase. Deaerators, additives that will remove entrained air from the furnish, will increase dry content of the fibre cake and thereby reduce the carry over of water over the wash filter. This will lead to an overall downstream reduction in colloidal material.

Recycling of water will accumulate the colloidal material and, in consequence, the higher the closeness of the water circuit, the higher the concentration of colloidal material.

Retention of Colloidal Material

Papermaker's alum is sometimes used for pitch control as alum functions by fixing the colloidal pitch particles onto the fibres, removing these from the system. The conditions (acid pH, etc.) by which this method can be utilised are the same as those required for rosin sizing.

Products used for filler and fines retention will partly retain organic colloidal material, although such products are not specific for colloidal material and not very efficient for the retention of colloidal material. More specific products, usually water-soluble highly charged cationic polymers of a molecular weight lower (less than 100,000 g/mole) than those used for fines and filler retention, are available to specifically retain colloidal organic material during sheet formation. It is believed that these products are able to bind the colloidal particles more strongly to the fibres, thereby minimising the potential deposit problems in the press and dry section due to an increase in colloidal material taken out with the paper. Such additives should be added continuously before the point of deposition. Typical addition rate is 50–500 g polymer per tonne paper. The compatibility with other additives such as retention/drainage agents in use should be ensured. Care should also be taken to ensure that the colloidal charge of the system is not reversed from anionic to cationic by the addition of a strongly cationic polymer.

One way of reducing pitch problems is to use a solid adsorbent. The organic material will either be adsorbed by the adsorbent or the adsorbent will, in a physical manner, coat the organic colloids. The use of a solid adsorbent can be seen to reduce the content of organic material available for deposition. The only commonly used adsorbent is micronised talc with small particle size distribution. Using this approach of organic deposit control, it is important to use an efficient quantity otherwise talc will increase the content of material by being incorporated into the organic deposits. What comprises an effective quantity depends on the content of organic material. An addition of 5–15 kg/t is typical for the industry today.

Minimising the Depositability

Depositability of the organic material is obviously required for a deposit to be formed. It is also easy to conceptually understand the relevance of this parameter. Depositability can be seen as the tackiness of the material. Consider a situation where the colloidal organic material is present in a certain content in the system. With a low viscosity this material has a low viscosity, and is not really tacky, so deposition of this material is not going to be extensive. At high viscosity, on the other hand, a material is brittle in nature so this material will not readily form deposits either. However, if the viscosity is in between these two extremes, and of a tacky nature, deposition can be expected to be extensive.

The tackiness of the organic material will depend, like the other three main parameters, on a number of factors: chemical composition of the organic material; and temperature of the system being the two most important. An increase in temperature will always reduce the viscosity of an organic material. It is a general experience in the industry that problems with organic deposition increase at increasing system temperature.

The viscosity of the organic material will also be determined by the chemical composition of the material. Rosin acids (having a high melting point) will increase the viscosity, while unsaturated fatty acids (having a much lower melting point) will decrease the viscosity. It is a general experience that fresh wood has a more troublesome pitch than stored wood, and that storage of chips will reduce pitch problems. During this process the wood extractives are modified by oxidation which increases the viscosity of the wood extractives and so the deposition is reduced. A further modification of pitch on storage is evaporation of terpenes and other volatile substances. Removal of such substances, of almost solvent nature, will also increase the viscosity.

Also pH and water hardness can influence the viscosity of depositable material as the fatty acids may be converted into calcium soaps or other soaps of divalent and trivalent ions. Increase in pH (up to approximately pH 9.0) and increasing water hardness will increase the formation of metal soaps of fatty acid. The formation of metal soaps of fatty acid will increase the viscosity of the organic material.

A deposit may also be formed when two components, each of them not forming a deposit, are combined within the papermaking system. The combined material can then obtain the tackiness required for deposition; a tackiness none of the components alone have.

Today no chemical additives are used to modify the viscosity of the organic material of natural origin (wood pitch) to reduce deposition. Water-insoluble solvents, that would reduce the viscosity, can conceivably be used reducing the viscosity of the resin. The use of kerosene for pitch control, a method used half a century ago, worked by this mechanism.

Maximising the Colloidal Stability

In the pulping process not only organic water-insoluble material is released from the wood but also water-soluble organic components. These organic water-soluble components of polymeric material comprising hemicelluloses, lignins and low molecular weight acids (e.g. acetic acid) also have some anionicity. It has been shown that some of these water-soluble components can stabilise colloidal organic material preventing its deposition. Figure 2.5 depicts how the water-soluble components adsorb onto the surface of the colloidal material. The presence of these natural colloidal stabilisers is the reason why papermaking using thermomechanical pulp (TMP) and other mechanical fibres (where the content of natural stabilisers is high) is not associated with worse pitch problems than papermaking using chemical fibres having significantly lower content of pitch.

The water-soluble organic material is often referred to as 'anionic trash' as it impairs the performance of all chemical additives and deteriorates the quality of the

Fig. 2.5 Colloidal particles stabilised with water-soluble material adsorbed onto the surface

paper. However, 'anionic trash' does have the very positive function of stabilising colloidal material. Removal of the stabilisers without removal of the colloidal material would trigger an extensive deposition of this colloidal material.

When waste paper is used as the fibre source, depositable organic material (adhesives, hot-melts, etc.) gets into the system without the simultaneous addition of natural stabilisers. As a result deposition of this material is often extensive. Deposition of organic matter will occur when the content of organic water-soluble stabilisers does not balance the content of colloidal depositable material. The balance can be restored by the addition of synthetic stabilisers. Synthetic stabilisers used for deposition control are always organic polymers of relatively low molecular weight (below 50,000 g/mole), Polymers used are of anionic, non-ionic or even cationic nature, and should be added continuously before the point of deposition. Typical addition rate is 50–500 g polymer per tonne of paper. Care should be taken to ensure that the addition of this type of product is compatible with other additives such as the retention agent. One important point of consideration is charge of the system, and it should be ensured that the charge of the system (either determined as Z-potential or colloidal charge) is within a range where good runnability is experienced. This range differs from one machine to another, but is usually slightly anionic.

Minimising the Surface Affinity

Deposition requires not only an organic material, but also a surface. The nature of the surface is quite important, and over the years the industry has learnt to avoid certain materials onto which organic deposits would quickly build up. Deposition of organic matter on a surface can be prevented/reduced not only by treating the depositable material using any of the processes described above to treat the furnish, but by modifying the surface onto which the deposit is formed also. Certain surfaces (like wires, felts and rolls) are particularly prone to deposition. Surfaces can be modified permanently to minimise deposition as in the polytetrafluouroethylene (PTFE) coating of drying cylinders.

Effective methods of continuous treatment are also available. Addition of a chemical product to the furnish always involves a considerable dilution (as the volume of water used in the industry is vast), but treatment directly onto a surface by spray application can be much more concentrated. For this reason it is often more cost-effective if the problem of deposition is defined to a limited part of the papermaking process. The most common method comprises continuous spraying of a water-soluble cationic polymer onto the surface treated. This cationic polymer will then react with anionic water-soluble macromolecules in the furnish (anionic trash) to form a coating. This coating, which is very thin and flexible, is sacrificial, and anything that deposits on this sacrificial coating will be removed from the surface as the deposit is redissolved. Often the polymer is formulated with other materials (like surfactants) to enhance the performance. For centre press rolls spraying a release wax emulsion seems to be the most successful application.

All the above methods are used in the industry to minimise organic deposition. The best result is often achieved by minimising the contribution of a deposit using a combination of these methods. However, there is no single solution that will provide the complete remedy to all deposition of organic material. Deposition of additives (defoamer, sizing agent, etc.) can often be prevented by avoiding excessive use, by changing the product or changing the addition point.

2.2.3 Inorganic Deposits

Deposition of inorganic material can be rather troublesome. Such deposits can grow to considerably thickness, as exemplified by the picture below (Fig. 2.6) taken at a big Kraft pulp mill located in a Nordic country.

Formation of inorganic deposits can take place in principally three different ways:

1. An inorganic deposit can be formed by precipitation onto a surface of a salt from water-soluble ions in the water. Such a deposit is called scale and does not require the presence of other components like organic material or microbiological slime for its formation.
2. An inorganic material, like filler, can be incorporated into an organic deposit or microbiological deposit. In these cases the organic material or the microbiological material (slime) will act as a binder, and without this binder the inorganic material would not normally deposit.
3. Inorganic filler particles can be mechanically trapped in a papermaking felt. In this case there is no absolute need for a binder material although its presence often aggravates the filling of felts by such particles.

Fig. 2.6 Scale deposit

These different modes of deposit formation will be dealt with individually.

2.2.3.1 Inorganic Scale Deposits

Formation of the scale nature of inorganic deposits is caused by precipitation of a salt from the water phase of the system. This form of deposit, called scale, is the most common form of inorganic deposits. There are several different types of scales found in the paper industry. Examples are calcium carbonate, calcium sulphate, barium sulphate, aluminium hydroxide and calcium oxalate. All these precipitates are formed because the water solubility of the respective ion pairs is exceeded.

$$\textit{Conc of anion} + \textit{conc of cation} = \textit{conc of precipitated salt}$$

The exact nature of scale is best determined by FTIR or X-ray spectroscopy. Anions and the cations respectively in the water phase can be determined through methods such as atomic adsorption (for cations) and ion chromatography (for anions). Unlike microbiological and organic deposits, determining the exact chemical composition of the scale is critical to establishing the best approach for reduction of scale formation.

Calcium Carbonate

Calcium carbonate scale is the most common of these three forms of inorganic deposits. It has the appearance of a hard coating on the surfaces of the paper machine or a pulp system, usually discoloured by incorporation of small quantities of other material such as lignin.

Calcium carbonate scale formation increases with increasing salt concentration and increasing temperature at a pH above 8.0. Below pH 8.0 the salt is quite water-soluble. It is frequently found on neutral sized fine paper machines located in hard water areas, where conditions for scale formation of this type are favourable, since the pH on such a machine is alkaline and the concentration of dissolved calcium and carbonate ions is high. On a paper machine it is not usually detrimental to paper production in itself unless it becomes extensive. However, it does provide an excellent surface for bacterial growth and deposition of organic material and for this reason should be prevented.

Calcium carbonate scale is even more commonplace in Kraft pulp mill systems such as heat exchangers, where, compared to paper machines, it often forms a thick deposit leading to loss of performance of the heat exchanger. Calcium comes into the system with either the wood or from poor recovery liquor clarification and carbonate is formed from degradation of sugars in a side reaction during pulping conditions. Dissolution of filler (calcium carbonate) will increase the concentration of both calcium and carbonate in the system.

Calcium Oxalate

Calcium oxalate has recently become a common scale due to increased closure of water circulations. This type of deposit may occur on paper machines using NSSC pulp, where the oxalate ion is formed in the NSSC pulping process as a degradation product of glucose. It is also becoming a common scale in highly closed bleach plants and TMP plants.

Barium Sulphate

Barium sulphate is also becoming a rather common scale, particularly in closed pulp mills. When it occurs, the problem of its formation is extremely troublesome as it is a salt with very low solubility. Because barium sulphate is a salt of such low solubility, barium sulphate scale may form even if barium concentration, relative to other cations, is low, particularly when concentration in the system of sulphate ions is high. Barium is dissolved from the wood in the digester. The sulphate ion comes into the system through additives such as alum, and high sulphate ion concentration may develop as a result of accumulation of the sulphate ion from papermaker's alum (used in the rosin sizing process) or from sulphuric acid used for pH control. A particular problem is Barium sulphate deposition in the headbox of Kraft liner machines.

Calcium Sulphate

Scale of calcium sulphate may appear in TMP refiners and sulphite mills, particularly cooking liquor evaporation plants. It will not develop on paper machines as its water solubility is very high at this temperature relative to other types of scale-forming salts.

Silicate Scale

Silicate scales may be formed in pulp mills and in de-inking units, but are less common on paper machines. However, it is somewhat more common in recovery loops of pulp mills. It is a rather common scale in Asia, where content of silicate may be high in the fibre source, particularly bagasse.

Aluminium Hydroxide

Alum is an additive used for many purposes in the paper industry. Examples of reasons for use of alum is rosin size retention, pH control, drainage increase, charge neutralisation and pitch deposition control. Alum can form several different species of aluminium depending on the system pH and the most useful form in papermaking is $AI_8(OH)_{12}$. This form exists only at pH 4.5–5.2. Insoluble aluminium hydroxide $AI(OH)_3$, is a form predominantly at pH 5.0–9.0. The use of alum in this pH range is often associated with deposition of aluminium hydroxide; a gelatinous type of deposit caused by hydration of the deposited aluminium hydroxide. Systems with natural hard water and/or a high degree of closure are most susceptible to this slime-like deposit and the best method to eliminate the formation of aluminium hydroxide deposit is to reduce the use of alum and have a good pH control.

2.2.3.2 Methods to Control Deposition

Sand blasting and high-pressure cleaning can be used to remove deposits already formed. Few mechanical methods to prevent scale formation exists, other than opening up the system to bleed out the dissolved salts to prevent accumulation above the solubility of the salt.

Deposits of all these types of scales can be removed using chemical cleaning additives. This method of removal is greatly superior to the alternative mechanical method of removal, such as sand blasting. Calcium carbonate and aluminium hydroxide scales are easily removed with acids while calcium oxalate scale and Barium sulphate scale are more difficult requiring specialised cleaning compositions.

Methods are also available that will effectively prevent formation of scale. Through the addition of chelants (EDTA, DTPA, etc.) it is technically possible to prevent scale formation. However, this is not a commercially viable process as chelants only operate on a stoichiometric basis, and the metal ion concentration is far too high for such an approach to be cost-effective.

Useful methods involve the use of low molecular weight polymers and phosphonates capable of interfering with the crystallisation process. These low molecular weight polymers (usually carboxylated or sulphonated giving the polymers a strong anionic charge) are very specific in their action. This is why determination of chemical composition is critical before a treatment programme is defined. The pictures below (Fig. 2.7) show the difference between scale formed with and without a scale inhibitor.

Prevention of calcium carbonate scale is relatively easy to accomplish usually requiring polymer/phosphonate concentration in the ppm range only. Prevention of calcium oxalate scale and barium sulphate scale requires higher (sometimes

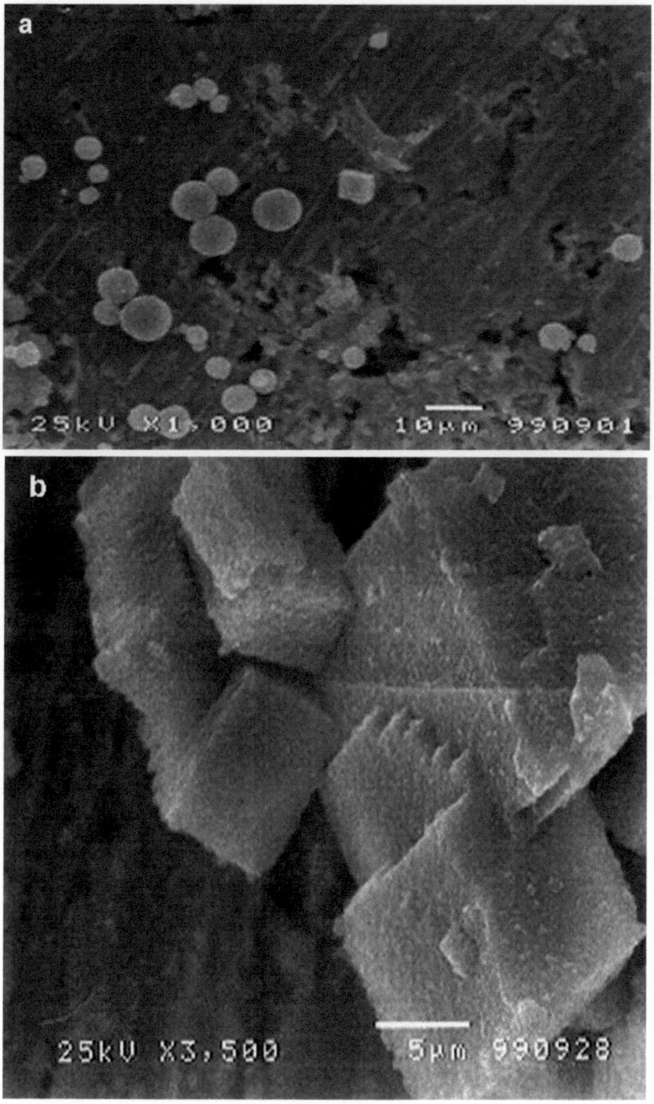

Fig. 2.7 Scale formed with and without scale inhibitor respectively

considerably higher) treatment concentrations. The product should always be added continuously to achieve the best effect regardless of the nature of the scale.

Establishing the most cost-effective treatment programme requires an extensive analysis of the entire system with regard to temperatures, water flows, salt concentrations, etc. Companies specialising in providing this service and providing the appropriate chemical additives are available.

2.2.3.3 Inorganic Filler Accumulated in Felts

A common problem in the paper industry producing grades containing filler, or when using recycled paper containing filler, is the mechanical entrapment of filler particles in press felts. Determination of the ash content in the felt and the subsequent analysis of the ash will determine the extent of this problem. Plugging of felts with inorganic filler particles is possible without the simultaneous deposition of an organic deposit in the felt acting as a binder. However, deposition of an organic material in the felt will aggravate the deposition of inorganic filler.

A few different approaches exist to minimise/eliminate this problem. To some degree this problem can be lessened in the design of the felts, where more open felts will cause less problems, and by eliminating the formation of an organic binder deposit in the felt by an appropriate method (see deposition of organic material) the problem can often be significantly reduced. An alternative useful approach, if the filler is calcium carbonate, is to continuously apply a treatment comprising an acid (sulphamic or sulphuric) that will dissolve the calcium carbonate. In this approach care should be taken to avoid a too low pH as this will burn the felt.

2.2.3.4 Incorporation of Filler in an Organic or Microbiological Deposit

Filler particles may well constitute the main component by weight in an organic or microbiological deposit, but are not in this case the cause of the deposit. Such deposits are best treated by minimising deposition of the binder material using methods to reduce deposition of organic material or microbiological slime depending on what the binder is. Usually the presence of such deposits is an indication of poor filler retention. Their formation can also be reduced by increasing the filler retention, which will reduce the filler content in the system. Increase in filler retention will not completely eliminate such deposits, but may reduce the extent considerably.

2.2.4 Deposits of Water-Soluble Components

In a pulping process not only water-insoluble lipophilic components are released from the fibres but also water-soluble hydrophilic components. This class of compounds contain hemicelluloses, lignins and low molecular weight organic acids and often the content of the water-soluble components exceeds the content of water-insoluble components. Further, bleaching will increase the release of water-soluble components from fibres particularly at alkaline conditions. Water-soluble hydrophilic extractives are easy to remove by washing, however for some paper grades, the pulp is poorly washed. TMP for newsprint contains approximately ten times more water-soluble hydrophilic components than water-insoluble extractives.

Despite the considerable quantity of these contaminants they are never found as components of deposits, and deposits of water-soluble components do not exist in the true sense. However, wet strength agent (that is water-soluble) may polymerise to form a rigid gelatinous network. This may happen in felts on paper machines using a wet strength agent and the permeability of the felt will be reduced. Although strictly speaking not a deposit, polymerised wet strength agent in felts leads to the same problems as a true deposit would.

Polymerisation of a wet strength agent in a felt can be prevented by changing the conditions in the felt required for polymerisation. Polyamidoamine-type wet strength agents require an alkaline pH for polymerisation. For this reason all machines using this wet strength agent run at neutral to alkaline pH. Polymerisation of this wet strength agent can be prevented in felts by the continuous application of an acid to the felt reducing the pH, thereby preventing plugging.

Chapter 3
Retention and Drainage

Paul Cutts

3.1 Introduction

As the paper and board industry continues to increase the speed and output of paper machines, the demands on retention and drainage aids are increasing greatly. Increased shear forces at the headbox and on the wire and ever shorter dewatering zones can be incompatible with the need for good runnability, high quality and good formation.

Increased water system closure and increased water temperatures are also altering the environment within which the retention and drainage aid has to function, requiring an alternative approach in many cases.

There is a large range of retention and drainage products available from many suppliers, which can be used alone or in combination as retention systems. These products and systems are identified chemically, their use explained and guidance given on the application, testing and suitability of the various systems.

The paper machine and system chemistry factors which can affect the retention and drainage aid performance are discussed along with some troubleshooting guidelines and examples.

The successful application of an effective retention and drainage aid chemical programme is becoming ever more complicated. New paper grades are being developed all the time, water usage is decreasing and system closure has an increasing impact on the products and systems used. More than ever, control and consistency of the papermaking process is vital as good runnability and high paper machine efficiency are essential to profitable papermaking.

What does a papermaker expect from a retention and drainage aid?

1. Retention – increased first-pass retention of fillers and fines, increased ash content of the sheet, reduced losses to effluent
2. Dewatering – improved wire section, press section and dryer section water removal
3. Formation – acceptable formation, better formation with lower headbox solids

P. Cutts
Solution Application Specialist, Ciba Specialty Chemicals, Cleckheaton Road, Low Moor, Bradford BD12 OJZ, W. Yorkshire, UK

I. Thorn and C.O. Au (eds.), *Applications of Wet-End Paper Chemistry*,
DOI 10.1007/978-1-4020-6038-0_3, © Springer Science+Business Media B.V. 2009

4. Production rate – improvement from faster dewatering rates
5. Production efficiency – reduced raw material costs, reduced drying cost, better runnability, lower overall chemical costs and lower effluent treatment costs
6. Additive efficiency – better retention and lower usage of size, dyes and other effect chemicals; reduced need for defoamers and biocide due to lower loading in the white water
7. Consistent performance over a range of grades and conditions (see Figs. 3.1 and 3.2.)

Papermaking is becoming increasingly split into two main areas: large, fast machines making a small number of grades and basis weight products, and smaller machines which are more flexible, varying grades, colours, sizing or basis weight every few hours.

The two areas present different challenges for retention and drainage, especially where optimum performance for a particular grade may need to be reduced for the sake of consistency across all grades.

For the supplier, having a wide range of retention and drainage aid products and systems is essential as everything starts with the needs of the papermaker, the products made and the papermaking machinery used.

Once a system has been adopted by the papermaker, the supplier has also to provide or ensure reliable product supply, reliable make-up equipment, reliable product dosing, ongoing technical support and troubleshooting services. Failure in any of these key areas will compromise the retention and drainage system chosen.

Regular reviews and reappraisals of the retention and drainage system are essential to maintain the high level of performance required by the papermaker, as changes to the pulps, fillers, effect chemicals, water usage and machinery can all have a significant effect on the system efficiency and may require an alternative product/ system to be implemented.

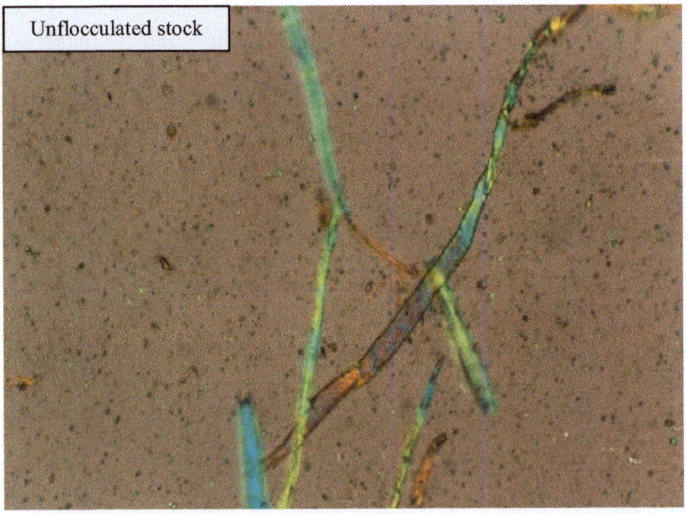

Fig. 3.1 Unflocculated stock – the fines and filler are apparent in the water phase

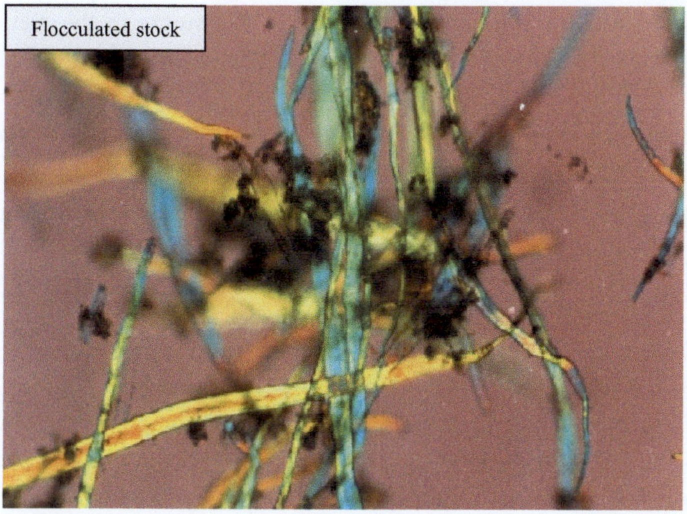

Fig. 3.2 After flocculation, they are mostly flocculated together and fixed to the fibres; the result is filler and fines retention and increased dewatering due to the floc structure formed

The text that follows is the result of 25 years of field experience and I have attempted to include all relevant products and systems available with equal weighting depending on their suitability for specific applications.

Any products/systems not included are either too specialist for inclusion in a general discussion or are unintended omissions on my part for which I apologize in advance.

3.2 Review of Chemistry

The following products form part of the considerations for the retention and drainage system which will eventually be used and should therefore be looked upon as components of the retention and drainage system – as any change in the product used or the dosage, application point, etc. can have a significant effect on the retention and drainage aid efficiency on the machine:

Aluminium species (alum, polyaluminium chloride, polyaluminium silica sulphate, etc.): used for charge control, pH control, as sizing mordants, broke treatment.

Synthetic coagulants, fixatives (polyamines, polydadmacs, polyethyleneimines polyvinylamines): used for charge control, as sizing mordants, broke treatment, stickies treatment.

Starches (wet-end starch – normally cationic, low, medium or high degree of substitution): used for strength primarily, with effects on system charge and as a key retention aid component in some systems. Where starch is a necessary component of the retention system, care needs to taken as to the best addition point to achieve the necessary balance between strength and retention.

Dispersants: usually anionic species used in the stock, in the wet-end or in coatings. These will affect the overall charge of the system and can cause problems when used intermittently.

The products normally considered as the effective retention and drainage aid additives are:

Polyacrylamide flocculants: low, medium or high molecular weight products of low, medium or high charge (cationic or anionic) and no charge (nonionic); available in a variety of physical forms – liquid dispersion, emulsion or solid grade

Polyethyleneimine: low molecular weight but structured cationic products available as liquids, generally these are more effective as drainage aids

Polyethylene oxide: very high molecular weight nonionic products, available as solid grades or dispersion/slurry

Microparticles: silica-based colloids, sol (milky colloid) or polysilica microgels (available as liquids or generated on-site), bentonites or other inorganic silicates available as powder or slurry

Micropolymers: polyacrylamide-based anionic or cationic structured polymers or cross-linked particles, available as emulsions or dispersions

Others: combination products which are mixtures or water-in-water emulsion products containing both coagulant and flocculant, available as liquids

As machine speeds have increased and wet-end system chemistry has become more complex, the demands on the retention and drainage aid have increased considerably. This has led to the development of an increasing number of retention and drainage aid systems which incorporate two, three or even four components selected from the above list. A very general representation of the time frame for this development is shown in Fig. 3.3.

The most common systems in use are:

- Dual component systems – coagulant with flocculant
- Microparticle systems – starch and or polymer with silica microparticle (in various orders of addition)
- Microparticle systems – polymer with bentonite (in various orders of addition)

Addition Points

A variety of addition points are used in practice, depending on availability and also the system which is to be applied.

Occasionally, two products can be mixed prior to addition or more than one product can be added at the same addition point – out of necessity or because laboratory testing has shown this to be the optimum addition strategy. The main addition points used for the retention and drainage aid components are illustrated in Fig. 3.4 – points 1–6.

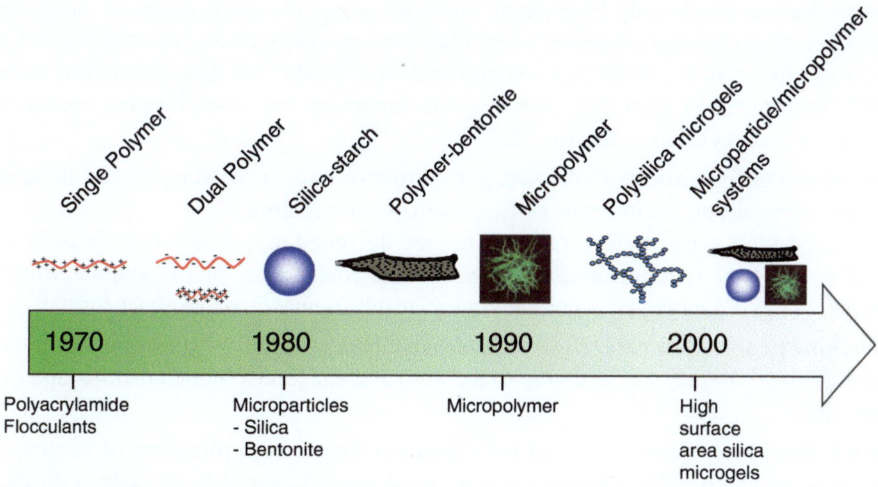

Fig. 3.3 Development of retention aids

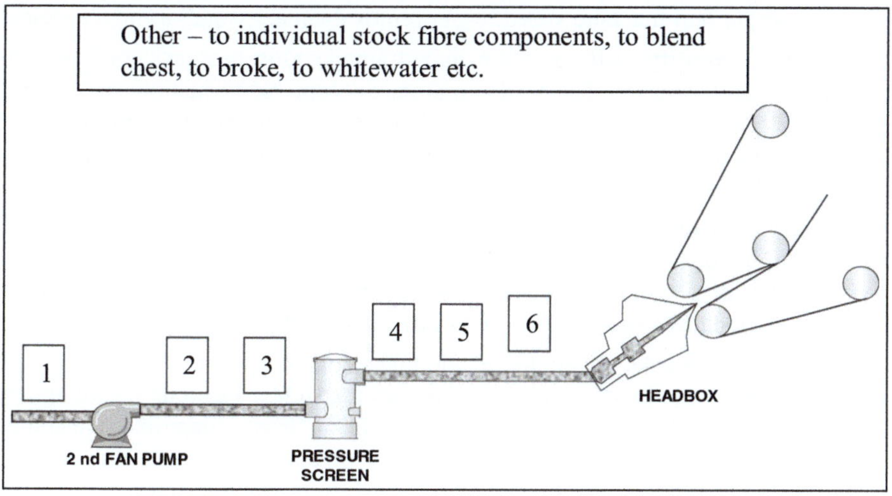

Fig. 3.4 Addition points

3.3 Basis of Selection

All of the above retention and drainage aid products and systems are in use currently. The system used depends on machine type, machine speed, paper/board type, production rate, requirements, etc.

The basic criteria used for selection of the most suitable retention and drainage aid product or system, include the following:

Paper/board produced: Fine paper machines generally have a greater need for retention than drainage, whereas board machines are the opposite. Newsprint/LWC/SC machines require both retention and drainage. Tissue machines are notoriously difficult in terms of applying retention aids due to the low basis weights produced and effects on cylinder adhesion etc.

Machine configuration: Twin wire, gap former or multi-ply vat machines all have different requirements in terms of free drainage and retention.
The latest high-speed twin wire machines are designed to dewater mechanically in a short free and vacuum dewatering zone. Implementation of a flocculant which increases the drainage rate significantly can result in unacceptable sheet formation.

Machine production rate: Bigger, faster machines generally require a more robust retention and drainage system due to the higher shear forces in the headbox and on the wire.

Stock approach flow: This can be a limiting factor in application of the most efficient products. The papermaking systems need to be able to cope with the changes necessary for implementation of an efficient retention and drainage aid system. Restrictions on pulper capacity, fan pump capacity etc. can reduce the suitability of the most effective system, requiring a less powerful system to be used, as can limited machine speed, water removal capacity and drying capacity.

Paper/board quality: All retention and drainage aids products and systems affect the sheet formation. The need to maintain acceptable formation can limit the type of system that can be applied and also the level of retention and drainage performance which can be sustained. As many trials fail due to an over-performing retention aid as fail due to an under-performing retention aid.

Variability of grades produced: Where a variety of different grades are produced, a balance must often be accepted between maximum performance grade by grade and the need for good performance across all grades. Flexible systems which can be adjusted grade by grade can be used most successfully.

Any ongoing problems: Owing to the importance of the retention and drainage aid system in day-to-day productivity, papermakers are rightly cautious regarding any changes to the system used. Where ongoing problems exist – deposits, runnability, productivity, quality, etc. the retention and drainage aid supplier will endeavour to determine which product or system can be applied to resolve the issue.

3.4 Methods of Evaluation

Free Drainage

Thin stock free drainage measurement, with and without addition of retention/drainage aid, can be carried out readily using a Schopper Riegler Beating and Freeness Tester or Canadian Standard Freeness Tester (see Fig. 3.5).

Fig. 3.5 Canadian Standard Freeness tester (Photo courtesy of Thwing-Albert Instrument Company)

Both can be used as they are, or better, modified by blocking the rear orifice to allow all the drained water to flow from the front opening. The draining water can be manually timed to fixed volumes (400 ml, 500 ml, 600 ml) or the water drained into a container on a balance so the weight of water can be logged directly onto a PC for graphical representation.

Faster draining stocks usually require 1,000 ml per test, while slow draining stocks can be evaluated with 500 ml per test.

As a fibre mat forms on the wire of the drainage tester, the drained water is unsuitable for retention measurements – it is very clean due to the filtration effect of the fibre mat. Turbidity measurements are sometimes carried out on the drained water where relevant.

Free drainage methods give a direct indication of which is the most substantive product or system and can give an indication of the optimum dosage, although product overdosing still gives good results.

As the standard testers may be too heavy/bulky to be readily portable, either a mill's own on-site equipment is used or supplier companies use their own lighter, portable equipment which operates on the same principle

Retention

The standard apparatus for retention determination is the Britt Dynamic Drainage Jar apparatus – usually referred to as the 'Britt Jar' (see Fig. 3.6). It has the benefit of being easily portable. The principle is always the same – draining thin stock while stirring to collect water draining through clean wire (no fibre mat forms) – the drained water being comparable to machine white water. This can then be used to determine the single pass retention for each test and also single pass ash retention.

$$\text{Total first-pass retention} = \frac{(\%\text{thin stock solids} - \%\text{whitewater solids})}{\%\,\text{thin stock solids}} \times 100\%$$

Fig. 3.6 Britt Dynamic Drainage Jar (Photo courtesy of Paper Research Materials Inc.)

$$\text{Ash retention} = \frac{(\text{wt. ash in 100 ml thin stock} - \text{wt ash in 100 ml whitewater})}{\text{wt. ash in 100 ml thin stock}} \times 100\%$$

Variables are

1. Stirrer speed – the minimum speed used is usually that which keeps the wire clean and does not allow a fibre mat to form. Higher speeds can be used when endeavouring to match the retention achieved on a particular paper machine (on the original Britt Jars stirrer torque can also be varied).
2. Wire used – a number of wires with different mesh sizes are available. When a piece of machine wire or a larger hole size Britt screen is used, lower retention measurements, more realistic in terms of on-machine results are achieved.

A typical testing method would be to take 500 ml stock in a measuring cylinder, add retention aid(s), mix by inverting the cylinder or mix on a separate stirrer, start the Britt stirrer, pour in the treated stock. After 5 seconds, collect the draining water for 20 seconds. The volume drained in 20 seconds should always be the same – if not, the wire is blocked. One of the common procedural errors is to add chemicals directly to the stock in the Britt jar. This can cause the wire to become blocked and should be avoided.

Vacuum Drainage

Various systems are available which determine the drainage of stock under vacuum. A portion of stock is treated as required and poured into a retention tester fitted with machine wire to which vacuum can be applied or into vacuum filtration equipment (Buchner, Hartley) onto a filter paper.

These tests, while often difficult to set up, can show the effects of the different types of flocculation which can be produced. In terms of the speed of drainage under vacuum, products will tend to be ranked in the same order as in the free drainage tests, but the differences occur in recording of the final vacuum through the formed pad/sheet. This indicates the relative formation/floc structure. The pad/sheet can then be weighed wet and dried to determine the moisture content of the formed pad/sheet. The best formation pad/sheet will have the lowest moisture content.

Using a machine wire, retention will be variable, while with a filter paper, retention will be constant – this will affect the results obtained.

Combined Systems

Devices which can be used for drainage and retention testing, which dose chemicals in a predetermined order and mix to a standard regime, are available (see Fig. 3.7).

They have the advantage of feeding results directly to a PC.

Fig. 3.7 Mutek DFR-04 (Photo courtesy of BTG). 'The **Mütek DFR-04 Drainage Retention Freeness tester** is the first laboratory unit that automatically detects the retention of fibres, fines and fillers. Within a few minutes it can be changed for drainage as well as freeness measurements. Advanced hardware and software allows simulation of paper machine conditions and evaluation of drainage and retention aids'

While laboratory tests allow selection of the optimum products and give an indication of dosage(s) required, they are only a guideline for machine trials. Final system optimisation and sometimes final product comparison can only be done on the machine.

3.5 Limiting Factors and Detrimental Substances

3.5.1 Stock Approach Flow

Pulper capacity – increased machine speeds may be possible, due to improved retention and drainage performance, but pulper capacity must be taken into account – can the pulper supply an extra X% of pulp?

Fan pump capacity – the fan pump can be a limiting factor in some mills where the machine is capable of higher speeds or more water is needed on the wire to reduce stock consistency.

3.5.2 Wet-End – The Biggest Problem Is Variability

As the retention and drainage aid is the last chemical addition to the papermaking system, it generally comes under suspicion whenever retention or drainage effect is reduced. There are however, many factors which can influence the performance of the retention and drainage system and depending on which type of retention/drainage system is in use it will be affected to a lesser or greater degree by fluctuations in wet-end chemistry:

Water quality – 99% or so of the thin stock is water. Water quality in terms of pH, temperature, solids, colloidal and dissolved materials has a significant impact on the wet-end chemicals and the performance of all additives. Changes in freshwater source and quality need to be monitored carefully as well as any change in system water closure.

Some changes in water quality, e.g. hardness and colour can be seen seasonally and these can be severe enough to compromise the whole chemical balance of the wet-end. Frequent monitoring is required in mills with sensitivity to these effects and also where the water used to prepare chemical additives varies in quality.

Good quality water is essential for the make-up and dilution of polymers and other products used for retention and drainage. The quality of the water used is capable of causing problems in polymer/product make-up, storage and also on-machine. The presence of solids in the water can lead to deposits; the presence of high calcium hardness can lead to deposits and also reduce performance of the product. Residual materials such as chlorine or alum must be avoided.

The most common cause of deposits in polymer/product preparation, filtration and dosing equipment is bacteria. Waterborne bacteria can thrive in make-up systems, causing slimy or pseudoplastic deposits to form. Once this has been identified, the systems need to be thoroughly cleaned, pipe work replaced and a treatment with biocide implemented. Care needs to be taken with the choice of biocide, it must be fully water soluble and compatible with the polymer/product otherwise reduced performance will result.

Filler content – where filler is used, it is often the single biggest factor in demand for retention aid. Filler has a high surface area compared to fibre. Increased filler levels will require increased retention aid dosages as well as increased levels of other effect chemicals – size, etc. Fresh filler and broke filler will have different anionic character and sometimes different surface area. Where these are variable, the system charge and system polymer demand can vary a great deal.

Conductivity – high conductivity shrinks the effective size of the flocculant polymer and reduces its efficiency. Generally, high molecular weight and lower charge polymers are effective in high conductivity systems. Some combination systems, e.g. bentonite followed by polyacrylamide, will function regardless of the conductivity of the system. Conductivity will increase due to increased water closure and this may necessitate a change in retention aid products.

pH – all quaternary cationic polymers (polyamines, polydadmacs, polyacrylamides, polyvinylamines) will function across a wide range of pH, but secondary and tertiary amines (polyethyleneimines) will have a higher charge at lower pH (<7). Anionic polymers react effectively across a wide range of pH.

Turbidity – increased turbidity in the stock can increase the demand for polymer as colloidal particles have a high surface area and absorb a disproportionately high amount of polymer. A cost-effective treatment can be application of a fixative to reduce the turbidity, followed by a polyacrylamide.

Particle charge – a qualitative measure of the overall charge in a system, but is better used to titrate the filtered stock with an opposite charged standard (polydadmac or sodium polyethylene sulphonate) to give a quantitative measure of the overall charge in the water phase of the wet-end. Colloidal and dissolved materials which affect the charge can vary and affect the performance of retention chemical systems. Of particular importance is to identify the causes of swings in charge demand, so the Particle Charge Detector (PCD) can be a useful troubleshooting tool. Generally, stock systems have an overall negative charge and may run into problems if the charge swings to cationic for any reason.

Zeta potential – a measure of the charge on the solids, as opposed to the PCD which measures charge in the water phase; a useful troubleshooting tool when retention aid performance changes.

Broke – variations in broke type and broke levels are one of the biggest recurring causes of variation in wet-end chemistry. Broke type can vary due to whether it is before or after a size press or coater, whether it is single-side-coated or double-side-coated, which recipe of coating mix has been used, etc. Broke treatment to reduce the impact of charge and/or white pitch is always recommended, preferably linked to in-line charge or less preferably, broke flow.

Addition points – most additions to a papermaking system are of charged species: anionic or cationic. Care is needed to ensure that all additives are added at the most effective addition point, allowing for mixing and adsorption effects, but also that no two additives which can react together are added close to each other (unless that effect is actually required). In some cases where a fixative and flocculant are being used, the optimum addition points need to be evaluated. Separate addition, far apart, may be best, but in some cases, premixing and simultaneous addition can be readily shown in the laboratory to be more effective.

Simple stab-ins or quills have been the standard addition points for many years, but a new generation of addition points is emerging. Metso's Retamix® system delivers the product into the stock flow at multiple points, while the Wetend Technologies Limited Trumpjet® systems allow additive dilution water to be replaced by clarified water or thin stock.

New versions of these technologies are now emerging, all aimed at improving the distribution of additives into the furnish.

Fibre types – changes in fibre type which cause a variation in wet-end chemistry also need to be monitored. Use of BCTMP in fine papers for increased bulk, for instance, can lead to increased demand for retention aid as well as for other effect chemicals, e.g. size, optical whitener. Variation in carry over of chemicals from de-inking plants can cause variable wet-end chemistry at the paper machine and where swings are significant this needs to be controlled. Addition of fixative may be necessary to reduce negative impacts.

Additives – all additives have the potential to affect the system response to retention and drainage aids – the main considerations are the additive charge and additive surface area, although some additives such as high cationic starch can also block sites on the fibres which the retention aid needs to use to function effectively.

Optical whiteners (OBA, FWA) are anionically charged and can have significant impact on the system charge and effectiveness of the retention and drainage aid.

3.5.3 Paper Machine

All aspects of the stock flow system and the paper or board machine equipment will have an effect on the retention and drainage aid.

Initial choice of the retention aid/system which can be used may depend on the capability of the fan pump flow or machine speed to increase; it may require adjustments to the foil drainage elements or vacuum boxes. Any changes to the system need to be evaluated for impact on retention and drainage aid – machine wires in particular. Changing a wire can increase or decrease first-pass retention on the machine by as much as 10% in extreme cases. Where drainage is readily achievable, using a more closed wire can be one route to increased retention.

3.5.4 The Sheet

Depending on the paper/board machine, the grade(s) produced and the chemistries involved, choice of retention/drainage aid may be limited by other factors – sheet formation requirements, sheet porosity specifications, sheet linting propensity, sheet bulk density, etc. Due care is required to maintain sheet quality at all times.

The most common cause of sheet defects is bacterial/fungal material. This can be reduced by determining the source – from an additive make-up system, building up in the stock system or the wet-end. Effective treatment needs to be installed to minimise the problem.

Defects such as stickies/pitch can be treated by an effective retention system – with an effective coagulant, pacified by using a bentonite as part of the retention system or separately pacified with talc or bentonite added early in the system.

3.6 Troubleshooting Experiences

When retention problems are reported, standard questions are asked:

What has changed?
Are all additives actually going into the system?
Is the problem continuous or intermittent?

Has the retention been manually checked?

In many instances, the problem can be easily resolved as problems are usually due to one of the above, rather than the retention aid itself. Retention systems are quite robust and will cope well with a certain level of variation in the system chemistry. Only when there is a drastic alteration in the additives to the system or the machinery will retention start to fail.

Example 1: Fine Paper mill-coated paper. Variations in retention performance were linked to broke usage. The higher the level of broke used, the better the retention. (Usually it would be expected to be the other way round, high broke causing problems.)

Investigation showed that the wet-end charge was overall cationic, an excess of cationic additives was being used. Medium to high levels of coated broke (with a high anionic charge) was essential to neutralise this excess cationic charge, allowing the retention aid to perform effectively. When low levels of broke were available, fresh filler with no added dispersant was used to make up the sheet ash to the required level, leaving the wet-end overcationised and the retention aid performance affected. Adjustment was made to the system additions to reduce the cationic charge input – of course, a dispersant addition to the fresh filler would have been equally effective although more costly.

Example 2: Fine Paper mill – coated paper. Variations in retention were recorded and the chemical suppliers were asked to try to determine the cause of the variation.

Broke levels did not correlate to the loss of retention, therefore the broke types were studied. Evaluation of the broke types for charge (PCD) and filler content showed that one particular broke was much more anionic than the others. When this broke was segregated and not used, the retention stayed at a constant level.

Eventually, production of the coated paper grade that produced the troublesome broke was moved to another mill within the group.

Example 3: White Top Liner mill – uncoated. The customer reported the presence of brown spots in the white top.

A check of the additions to the system revealed that an alum addition had been moved to a point in the stock flow opposite the bentonite addition of the retention system used. The alum and bentonite formed brown coagulated spots which showed up in the sheet, as well as reducing the efficiency of both additives.

The alum addition was moved further back in the stock system and the problem spots disappeared.

Acknowledgements The author wishes to thank Ciba Specialty Chemicals for allowing the use of illustrations contained in this chapter.

Chapter 4
Foam and Entrained Air Management: A Practical Perspective

Sal Mirza and Philip M. Hoekstra

4.1 Introduction

Paper machines and the papermaking process have undergone remarkable changes in terms of chemistry applications and design parameters. These trends increase the likelihood that the stock will foam and generate turbulence, thus allowing air to be drawn in. For example, higher machine speeds and gap former technology cause more air to be present in the white water, which could ultimately lead to foam problems if not managed correctly.

Surface foam and entrained air can indeed be a serious hindrance to operations and quality in a paper mill. This can be a major economic problem; recently it was estimated that the use of defoamers in the US paper industry alone in 2001 was over US$ 250 million and growing at over 5% annually. That is not the only cost of the problem - a much more significant impact to the papermaker would be the lost production and product quality problems that require the use of these products to control foam.

In papermaking systems, the gas in fibre suspensions is generally air (carbon dioxide can also be present depending on the process chemistry and conditions), which can exist in three forms (Matula and Kukkamaki, 2000):

- Free air
- Bound air
- Dissolved air

The least problematic of these three forms of air in papermaking fibre suspensions is dissolved air. Because air has limited solubility in water (~120 mg/l), any dissolved air is in solution and will cause no problems if it remains so. The combination of bound air and free air in a pulp suspension make up the total entrained air. If this entrained air could be viewed under a microscope, very small spheres of air bubbles dispersed into water would be observed (see Fig. 4.1).

S. Mirza
Assistant Professor, Midway College, Midway KY, USA 40347
e-mail: smirza@midway.edu

P.M. Hoekstra
Director Systems and Support, Buckman International
e-mail: pmhoekstra@buckman.com

I. Thorn and C.O. Au (eds.), *Applications of Wet-End Paper Chemistry*,
DOI 10.1007/978-1-4020-6038-0_4, © Springer Science+Business Media B.V. 2009

Fig. 4.1 Photograph depicting entrained bubbles in a liquid phase

Table 4.1 Problems caused by foam in pulp & paper mills

Foaming
Poor dewatering
Reduced washing efficiency
Reduced equipment capacity
Faulty instrumentation readings: influencing consistency measurement
Fines and fibres flotation
Reduced drainage efficiency
Pump cavitations, resulting in pump damage
Reduced mixing efficiency of pulp suspensions
Pinhole formation
Lower sheet strength, smoothness and bulk
Poor formation
Poor runnability: sites for pitch/dirt and other deposits
Increased microbiological activity
Black liquor carry-over
Chemical carry-over
Tank and filter overflows
Fibre loss through drains
Soda loss through drains
Compromising of system volumes, e.g. cleaner systems

Free air bubbles in the range of 70–100 μm in diameter tend to rise to the surface of the pulp suspension, resulting in the formation of surface foam. Bound or trapped air bubbles less than 70 μm are attached to the fibre or trapped within a fibre bundle or floc. As the entrained air bubbles rise, they may coalesce to form larger bubbles, subsequently bringing fibres and fines with them to the surface. This results in surface foam that contains fibres and fines.

Foam formation and entrapped air are problems that can adversely affect the efficiency of the pulp and papermaking process and thus negatively impact the quality of the end product. Table 4.1 summarizes many of the problems associated with air in pulp suspension (Pietikainen, 1992; Helle et al., 1999). Sheet breaks and pump cavitations can create runnability problems, including local pressure fluctuations (potential need for higher-capacity pumps) and deposition on the forming wire, press and dryer sections. In the forming section, entrained air hinders dewatering, prompting undesirable flocculation that negatively affects formation (Schwarz et al., 2002). Pinholes, high porosity, poor print quality, loss of strength, machine direction (MD) profile variations and poor formation are some of the noticeable effects on paper quality (Taber, 2003; Rauch and Burke, 2002). The combined net effects of the above problems lead to financial losses for the pulp and paper manufacturer (Helle and Paulapuro, 2004). The following list is a starting point identifying problems in the pulp and paper mill associated with air content.

4.1.1 Examples of Problems

As mentioned above, problems related to surface foam and/or entrained air can manifest in many different ways. Here are just a few specific examples from paper mills.

Newsprint: This example is from a mill producing standard newsprint from a furnish mix consisting of TMP and recycled fibre. The machine runs at a temperature of 45–55°C, and production is 600 t/day. Foam build-up is evident only in the wire pit and silo, but without adequate control the foam can grow to 15 feet in height. This foam overflows the wire pit and creates a housekeeping problem in the basement of the mill. Also, the wire pit has a mist removal system that will fill up with foam. At one point the duct in the mist removal system collapsed because of foam build-up (a serious safety problem).

Tissue: In a mill manufacturing primarily facial grades from bisulfite pine and Kraft hardwood, the primary cause of foam is the wet strength resin used in production. The negative impact of foam is twofold. First, heavy foam would fill the basement, where the electrical room is located, creating the potential for a major fire. Second, entrained air in the stock would increase the workload on the fan pump, reducing machine efficiency and production.

Wood-containing speciality grades: This example is from a fine paper machine producing 80 t/day of speciality grades, with basis weights ranging from 40 gsm to 110 gsm, at a pH range of 4.2–8.0, and furnish consisting of 80% recycled, 10% broke and 10% virgin. Some of the grades produced are envelope, carbon black, and deep dye grades. The problem from foam is lost production due to foam causing wet-end breaks, and reduced quality from spots on the sheet. Changing to an effective defoamer reduced lost time by 85%.

Paperboard: This mill produces board for tubes, cores, furniture backing and roll covering on a 160 t/day 7-vat cylinder-board machine. The system pH is neutral, and the temperature is 60°C. Water systems in this mill are nearly 100% closed. Additives used in the wet-end include alum, rosin size, cationic retention aid and various dyes. Poor foam control results in high entrained air measurements in the vats (>6%) and excessive surface foam to the point where the foam would overflow the vats. Quality is negatively affected by foam and entrained air: sheet crushing, blowing, poor plybond and at times foam spots on the backside of the board. The end result is off-quality board, poor machine production due to breaks and poor runnability in general.

Although the actual amount of entrained air that can lead to papermaking problems depends on the papermaking process, type of paper machine, grade, and furnish, even small quantities can be detrimental to paper properties and papermaking (Helle, 2000). For wood-free, wood-containing and speciality grades, the percentage of air content in the headbox can be as low as 0.1% but is generally in the range of 0.1–0.5% to 0.8% on the high side. In older packaging container and converting grade manufacturing, air is drawn into the process via open chests/tanks and worn seals and can be as high as 3–6% at the headbox.

One option to manage entrained air and surface foam is the use of chemical additives. In many cases a chemical approach is the only viable option. These additives are commonly referred to as defoamer or antifoams. Defoamers and antifoams are complex products comprised of many components and chemistries. Considerable progress has been made over the years in the technology of foam and entrained air management in the pulp and paper industry since the introduction of improved technologies as a defoaming aid.

The majority of the chapter is devoted to entrained air and foam management in the paper mill. However, foam causes problems in the pulp mill as well, impacting paper machine operation and paper quality. For example, air content can affect the efficiency of brownstock washing in pulping processes. Effective washing is not obtained because air (in the form of foam) compromises the venturi effect of the dropleg vacuum. Another effect is that air will act as a particle in the fibre mat, reducing drainage. This will impact drum speeds through impeded drainage, thereby slowing production rates. But more importantly from a paper machine perspective, this leads to poor washing and increased carry-over of detrimental substances. An effective defoamer washing aid at this stage of the pulping process provides cleaner pulps by promoting effective washing of wood extractives, which if carried forward can cause unnecessary additional chemical usage in the bleach plant or paper machine and could also cause severe deposition problems affecting both production processes and product quality.

This chapter presents information on what foam/entrained air is, its primary causes and effective strategies of control. In addition, it examines the choices of products available based on chemistry and application, and provides troubleshooting options for process engineers, suppliers and papermakers.

The following definitions provide a common understanding for terminology used in describing different types of foam, air problems in stock systems and chemical additive functions:

Surface foam: Dispersion of gas in an aqueous system at the surface

Entrained air: Dispersion of air in an aqueous system in the body of the system

Static foam: Old surface foam from which most of the liquid has drained/dried

Defoamer: Chemical additive to control foam and treat existing entrained air problems

Deaerator/antifoam: Chemical additive to prevent build-up of entrained air

EBS: Ethylenebis(stearamide) wax, very hydrophobic defoamer component

Silicone: The most hydrophobic component used in defoamer formulations, may also be identified as siloxane technology

Silica: Silica particle used in defoamer formulation; modified by attaching silicone to the surface to result in a hydrophobic particle

4.2 Foam Creation and Stabilisation

Every pulp and papermaking process has air in its pulp suspension. However, for the types of problems identified in Table 4.1 to occur, an additional component is required. Surface active agents, such as surfactants, must also be present in the system. As mills strive to meet environmental demands, reduce freshwater usage and lower production costs, system closure has resulted in increased levels of detrimental substances and higher stock temperatures.

One cause for paper machine foam problems is residual carry-over from the pulp mill or bleach plant. As much as possible the raw materials used should be as uncontaminated as possible. Black liquor is just one example of a contaminant that can cause foam on a paper machine. Another potential foam contributor is recycled fibre, used as part of a furnish. The recycled fibre contains all the ingredients used in paper manufacture and converting: coatings, sizing, inks, adhesives and binders, wood components, etc. that impact foam creation and stabilisation.

In addition, all of the wet-end additives encountered in a pulping and/or papermaking system can help to create and stabilise entrained air and surface foam. Additives such as starch, sizing agents, retention additives, wet strength additives, system charge modifiers, and de-inking chemicals are examples of wet-end additives which, if not applied appropriately, facilitate the creation and/or intensification of foam. Aspects for consideration include product selection, application point, dose rates and control strategy. It is clear from the preceding information that most of the additives that potentially cause foam are added deliberately to give specific properties to the paper or to improve the papermaking process. With this in mind, it is critical to balance the wet-end chemistry not only to optimize additive efficiency but also to reduce foaming tendency.

Additives increase foaming tendency, but also there are mechanical and process factors to take into consideration. Entrained air in stock systems can be attributed to mechanical entrapment of air in falling or turbulent stock, leaky pump glands or seals, open chests/tanks, and poor pipe design. For example, when liquid soap is added to water and agitated, it is observed that as increased energy is applied, increasing agitation, the level of foam increases. Applying this concept to the papermaking process and exacerbation of foaming problems it is important to consider stock mixing resulting in excessive agitation, cascading stock flows, and cavitations in pumps.

Stock temperature is certainly important as foam tends to be less stable at higher temperatures. This is one reason drainage often improves on a pulp or paper machine by operating at higher temperatures. For example, it is common for machines making unbleached Kraft liner board to operate at about 60°C with the benefit of increased production.

pH is another factor in foaming. Several foaming materials are affected by pH; for example, fatty acids that are a component of wood resins/pitch form soaps foam at higher pH levels, but these materials do not foam at low pH. If a furnish containing calcium carbonate is exposed to pH levels much below 6.5, a chemical reaction occurs that releases carbon dioxide, which appears as foam in the process.

In summary, process conditions such as pH and temperature significantly affect foam generation. Increased pulp mill chemical carry-over and the use of recycled fibre as well as high yield pulps will elevate the levels of detrimental substances in the papermaking system.

These detrimental substances are a source of surface active products that directly contribute to problems caused by air in the system, resulting in the creation of foam. The surfactant concentrates at the boundary of the water and air, resulting in a stabilising effect. These surface active substances (surfactants) consist of two distinct components: a hydrophobic component and a hydrophilic component. Figure 4.2 depicts a soap molecule, showing the hydrophobic and hydrophilic components. The hydrophobic (water repelling) component generally is a hydrocarbon chain while the hydrophilic (water loving) component is an ionized group or a water-soluble polar moiety.

As the concentration of these substances increases, there is a reduction in the surface tension of the pulp suspension resulting in high foaming capacity. The hydrophobic component will ordinarily align towards the centre of an air bubble. As a result, these air bubbles are composed of a three-layer film, as shown in Fig. 4.3, consisting of two air-water interfaces (a water layer is sandwiched between two layers of surfactant molecules). The surfactant molecules can make the bubble a very stable structure. The hydrophilic portions of the surfactants have a strong affinity for each other and for water. Therefore the viscosity of the water is increased, reducing the effect of gravity to cause the water to flow downward. The hydrophobic tails of the surfactant protect the structure from damage and also repel the surfaces of other bubbles nearby and thus reduce the tendency for the bubbles to coalesce. The result is a foam structure that can be very stable.

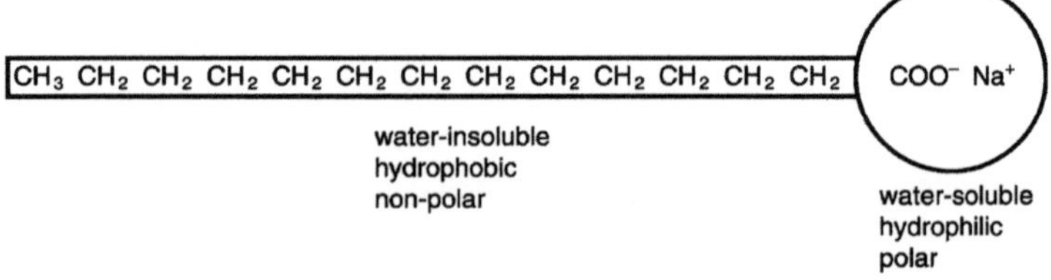

Fig. 4.2 Soap molecule showing the hydrophilic and hydrophobic ends

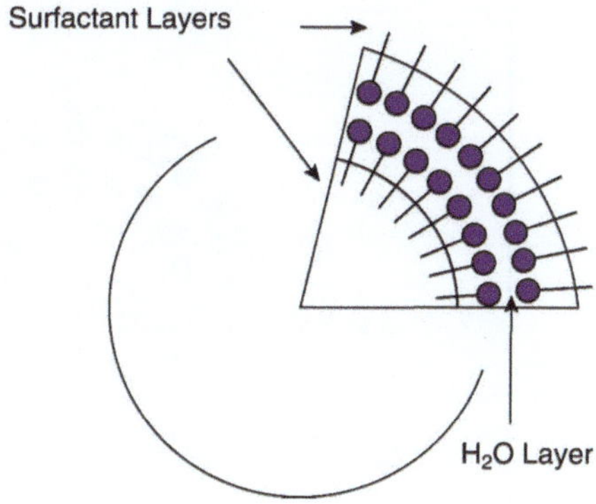

Fig. 4.3 Surfactant orientation to create a three-layer film

This orientation of the surfactant molecules around the air bubble creates a very thin film around the air bubble that can be strong enough to preserve the bubble. In the pulp and papermaking process, these types of bubbles can either appear in the water phase as entrained air or on the surface as foam (shown in Fig. 4.4).

Surface foam has a non-spherical shape that may be described as polyhedral. Compared with spherical bubbles, polyhedral foam has a larger volume of air in a relative small volume of water. It contains a large number of individual air cells separated by thin water layers (shown in Fig. 4.5).

Figure 4.6 shows the difference between the small bubbles in the depth of the water and the large foam cells at the surface. The bubbles will tend to rise, coalesce to form larger bubbles and collect near the surface, where due to drainage and coalescence of bubbles the foam is transformed into polyhedral or macro foam. Figure 4.7 shows bubbles coalescing.

Present in the papermaking process are air, water, and detrimental substances or surfactants that stabilise foam and entrained air. A fourth component that further enhances foam stabilisation is the presence of fines in the white-water system and papermaking process. Fines are finely divided particles that are often the result of the pulping or papermaking process, or they may be filler or pigment particles added to provide key paper properties. These particles tend to insert themselves between bubbles to interrupt the processes of foam or bubble destruction.

4.3 Foam Destruction

Although there are effective mechanical means to manage entrained air and surface foam, such as the use of deculators and resolving mechanical issues, the focus here will be on chemical means to destabilise and destroy surface foam and remove entrained air from the system.

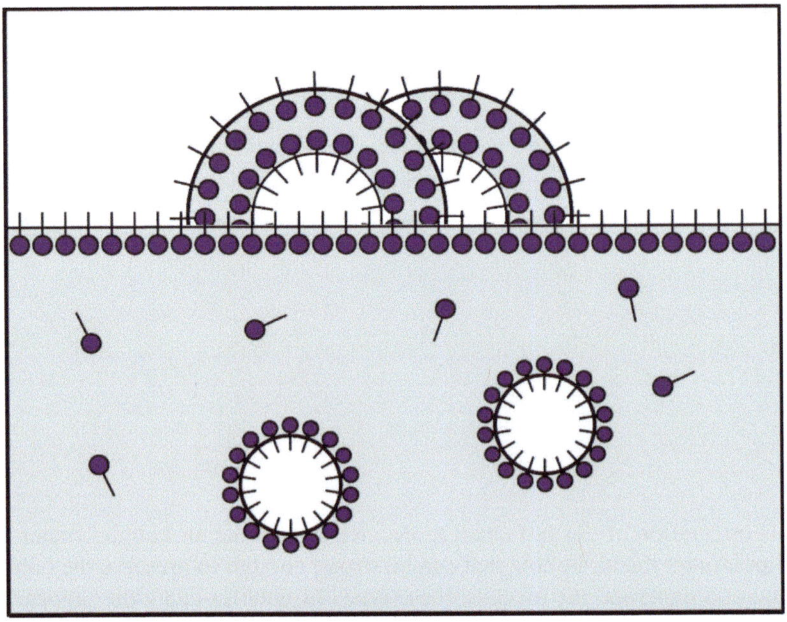

Fig. 4.4 Surfactant stabilised entrained air and surface foam

Fig. 4.5 The non-spherical nature of surface foam

Research conducted over the last several years and recently published literature have provided an understanding of the mechanism by which defoamers perform. Garrett (1993) contends that the oil in oil-based defoamers is present at the water-air interface causing bubble coalescence. He also states that the hydrophobic particles used in defoamers are lodged in the oil-water interface. Further work

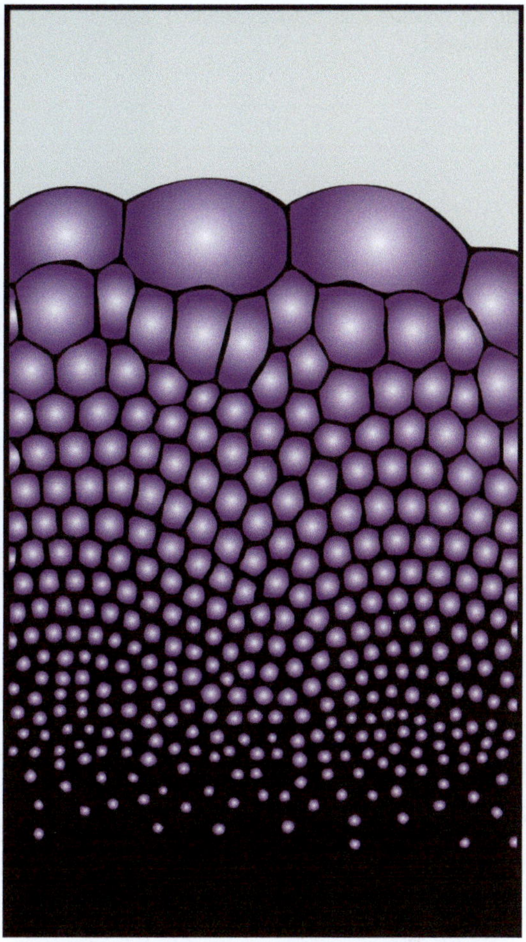

Fig. 4.6 Diagram depicting entrained air and the different bubbles sizes as they rise to cause surface foam

addressing the mechanism of oil-based defoamer products by Wang and colleagues (1999) have confirmed the work of Garrett and others who claim that hydrophobic particles in oil-based defoamers become lodged in the oil-water interface.

To act as an effective defoamer or deaerator, the surface tension of the product has to be lower than that of the foaming liquid. The product must disperse uniformly and quickly and not be soluble into the water phase. Other factors that contribute to a product's efficacy are:

- Particle or droplet size
- Concentration of emulsifiers in the formulation
- Formulation components, such as surfactant used
- Degree of hydrophobicity
- For defoamers containing particles (silicone, EBS wax, Fatty alcohol wax, silica):

Fig. 4.7 Photograph showing bubble coalescence

  - Size, surface area, and shape of the particle
  - Molecular weight and molecular weight distribution
- Process conditions such as pH, temperature, and conductivity

Of the large number of defoamer or antifoam products available, most will perform according to one of the following principles:

- Reduction of surface unity: All defoamers could be said to operate by disrupting the surface layer. Some defoamer chemistries are made of particles that migrate to the bubble surface and disrupt the regular packing of molecules at the water/air interface. Some particles may absorb surfactant molecules onto their surface. This reduces the surface cohesion. Examples are hydrophobic silica particles, EBS particles, or silicone particles.
- Spreading: The defoaming agent, because of its lower solubility, enters the foam lamella by spreading over the bubble surface. Foam-stabilising surfactants are repelled by the defoamer, which causes them to move towards a position in which surface tension is higher and where air bubble stabilisation is effected. The result is a weak area on the lamella and bubble ruptures. This is shown in Fig. 4.8.
- Particulate dewatering: An example of a particulate dewatering defoamer is one containing hydrophobic silica particles. The silica particles are coated with insoluble silicone films, resulting in particles that have such a low surface energy that they exert a dewatering action on the foam lamella.

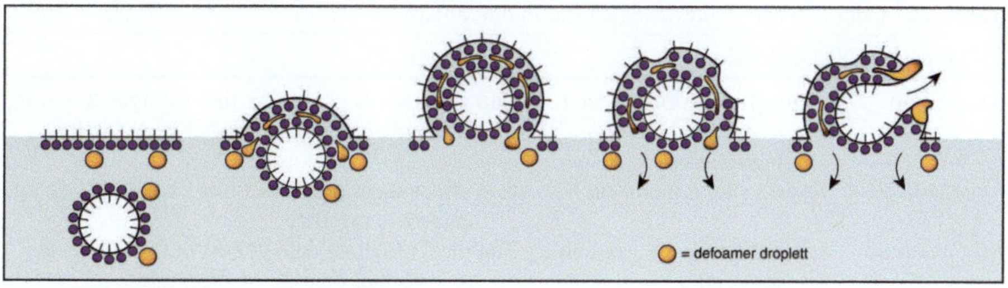

Fig. 4.8 Mechanism for bubble rupture

4.4 Product Type

Considerable progress has been made in foam and entrained air management in the pulp and paper industry since the introduction of petroleum as a defoaming aid. Because of the complex nature of the paper mill system and the fact that the chemically stabilized foams generally encountered are hard to break down mechanically, more mills are finding it not only helpful but necessary to use chemical defoamer formulations. These are generally emulsifiable combinations of hydrocarbons, alcohols, fatty acids, and various surface active agents, that frequently are specifically formulated for particular applications.

Table 4.2 lists several different common active ingredients used in defoamers in the pulp and paper industry.

To be effective the defoamer must spread quickly in the liquid, then destabilize the foam, and also maintain effectiveness for some time. The three key performance criteria are:

1. Spreading: Disperse quickly and interact with the foam and air bubbles
2. Knockdown: Immediate effect of foam knockdown and reduction in entrained air levels.
3. Staying power: Performance longevity or endurance to continue the high level of performance after the immediate knockdown as the product persists in the process

Traditionally, oil-based products have been used in the pulp mill as well as the paper mill. The oil-based defoamers can be generally described as dispersions of waxes and/or silicas in oil containing dispersants, emulsifiers, and other key components that modify the surface activity of either the products or the process liquors in which they may function.

There are problems associated with the use of petroleum-based defoamers. Volatility in supply and cost of the petroleum which is the base raw material is one concern. Another downside to the use of these types of defoamers becomes apparent as consumption rates are increased to their maximum. These products tend to complex with resins/pitch in the stock system, resulting in paper machine deposits and/or

Table 4.2 Common active ingredients used in defoamers

Component	Description	Comments
Mineral oil	Spreads on surface; used as a carrier for other ingredients	Can affect sheet properties like strength & sizing; environmental issues; source is petroleum
Vegetable oil	Works like mineral oil	Can affect sheet properties like strength & sizing; renewable resource
Silicone fluid	An extremely hydrophobic material	The most effective defoamers contain silicone, but it cannot be used on all grades.
Hydrophobic silica	A silica particle with silicone reacted onto the surface	To be used, this particulate must be dispersed in a carrier
Stearamide (EBS)	A very hydrophobic fatty-based material	Not often used in paper machines due to its propensity to cause deposits
Fatty alcohol	Particle, effective at moderate temperatures	Particle is dispersed in water-based formula; commonly used on paper machines
Alkoxylated surfactants	These are surfactants that destabilise foam	This is a very common product for controlling entrained air; a great variety of products exists
Silicone glycols	These are silicone-based surfactants	

paper defects. Therefore there has been a significant trend to move away from these oil-based products to oil-free formulations – water-based or concentrated.

To address machine performance and paper quality issues, alternative chemistries are continuously being evaluated and proven within the industry. Present technology includes products with chemistries such as synthetic alcohols, hydrophobic wax particulates, and synthetic esters. Chemical defoamer formulations containing emulsifiable combinations of hydrocarbons, alcohols, fatty acids, and various surface active agents are effective foam suppressors and deaerators.

Technology developments are moving in the direction of environmentally friendly water-based products to control surface foam and deaerate paper stock systems. Current technology allows formulation chemists to readily develop products that are capable of effectively combating both surface foam and entrained air through formulation modifications incorporating fatty alcohols in combination with other specific components.

Unique chemical properties of these new technologies include the following:

- Water-based
- Excellent spreading characteristics
- Non-toxic and chemically inert
- Effective at high temperatures in white-water systems
- Highly effective at lower dosages
- Environment-friendly
- Very effective in reducing entrained air levels
- Improved foam dispersing property
- Improved longevity in the system

4.5 Foam and Entrained Air Monitoring in a Mill

With reference to surface and static foam, monitoring and control are carried out with observations of foam build-up or levels in critical white-water silos or chests. Paper machine operators generally tend to make visual notes of the level of foam in a chest or silo, and if the level is above a known norm, then the defoamer addition rate would be changed.

Two methods are used to measure and monitor entrained air in a stock system: compression and ultrasonics. Many different types of mechanical compression apparatus have been used in the pulp and paper industry (Woodworth, 1990). The portable Entrained Gas Tester (EGT) is one device that is commonly used to profile stock systems by determining the level of entrained air in a stock at a particular sample point. The EGT operates on a mechanical compression technique based on Boyle's law of gases. The operation of the EGT is described in detail in the manufacturer's manual. Advantages of the portable EGT are that the test is quick to carry out and the test unit is compact and portable, allowing the user to conduct the test at-location. However, this test is very operator-sensitive and there can be high variability of data in addition to a time delay before results become available.

As papermaking processes have become more and more complex (both equipment and chemistries), the demand for 'real-time' data has been recognized as effective in providing timely and effective decision-making and problem-solving. To this end, wet-end online instrumentation has evolved, resulting in online units that are reliable and accurate: freeness, consistency, entrained air, drainage, charge and turbidity are some areas that have received attention.

Existing online methods used to measure entrained air in pulp and paper applications are based upon density, mechanical compressibility and ultrasonic velocity. The latter two have received the greatest acceptance with the industry. The compression method relies upon the principles of Boyle's Law, wherein the entrained air in the form of bubbles is measured by compressing the stock sample in a measuring cell. Ultrasonic entrained air units are mostly used for online measurement. The sonar measurement technique measures the volumetric percentage of entrained air in the stock sample. The basic principle of operation and measurement is that a transmitter emits a beam of sound waves through the process pipe to a receiver located on the opposite side, and the time of transmission is measured based on diminution and scattering of ultrasound-caused air bubbles in the pulp suspension. The amount of entrained air is determined by measuring the speed at which the waves move through the stock sample. The entrained air percentage is then calculated directly from the measured sound speed (Gysling and Loose, 2004).

Online measurement capabilities are increasingly targeted to identifying problems at an early stage of the papermaking process, thus allowing for a proactive approach to problem-solving. One of the benefits of online control of entrained air is to optimise the efficiency and application of defoamer additives. As an example, one paper mill through the use of online loop control to manage entrained air levels by controlling defoamer/deaerator application rates reduced overall programme consumption by as much as 37% (Watzig and Rauch, 2004).

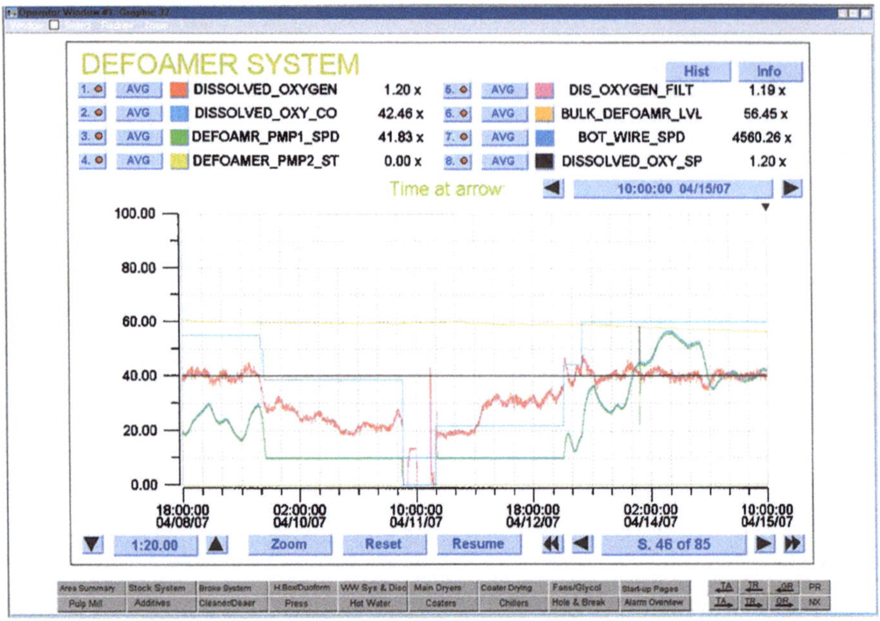

Fig. 4.9 Online entrained air management –effect of paper machine speed on entrained air levels

Online monitoring and control do provide additional benefits. The following examples are from a papermaking system producing a lightweight coated sheet using an online sonar-based entrained air measurement system. Figure 4.9 shows entrained air closed loop control with a set point of 1.2% entrained air. The left-hand vertical axis depicts the defoamer pump feed rate set point. The solid horizontal black line (at the 40.00 level) is the set point for entrained air: 1.2%. As the entrained air level dipped below the set point, the closed loop control system signals the defoamer feed pump to adjust product flow. From 9 April to 12 April the entrained air levels were below the set-point. Yet the feed pump remained at a setting of 10%, because this is the minimum set point for the feed pump to operate effectively. Then as the entrained air levels rose above the set point, the defoamer feed pump increased accordingly to maintain the entrained air levels at the set point.

The reduction in entrained air starting on 9 April was related to the decrease in paper machine speed. Conversely, as the paper machine increased speed starting on 13 April the entrained air levels increased as indicated by the defoamer pump feed rate setting increased.

Figure 4.10 shows the effect of maintaining a set-point of 1.2% for entrained air levels and locking the defoamer feed pump at 65% of the maximum flow possible (on 24 April). The effect observed was that the entrained air levels started to fluctuate rather than defoamer product flow. The wild swings observed on 22 and 23 April were identified as the periods when one of the paper machine deculator systems

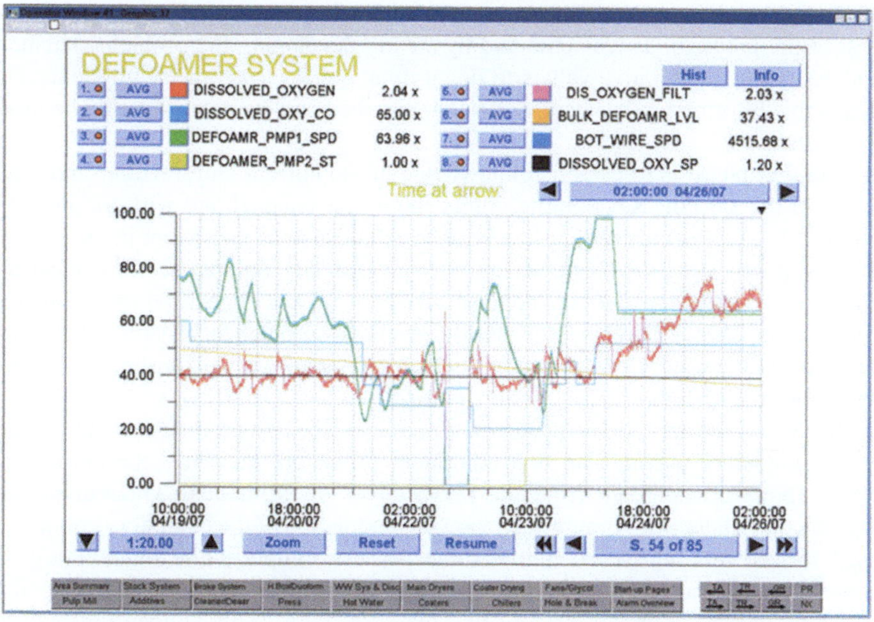

Fig. 4.10 Online entrained air management – impact of the deculator operation on entrained air levels

was in and out of operation. The swings in entrained air quickly set into motion a troubleshooting process identifying the deculator as the root cause.

Most online systems, when used in the capacity of control rather than monitoring, establish a set-point for entrained air at a particular place in the papermaking system and allow the defoamer addition rates to change. In other words, the entrained air levels as measured by the online analyser are used to adjust the dose rates of the defoamer/deaerator additives.

The online units are capable of measuring entrained air in stock samples with consistencies of 0.5–3.0%. Most units have a percent air measurement range of up to 8% by volume and a measuring accuracy of 0.01–0.03% by volume.

4.6 Application Strategies and Troubleshooting

The successful application of defoamer or deaerator programmes start with process knowledge and laboratory screenings of product chemistries to determine the most appropriate product(s) for the problem at hand. Each system will have specific problems and the final solution has to be tailored to each process. The overall guidelines in this section can be applied to gain knowledge and be applied towards troubleshooting, resulting in the identification and implementation of possible problem-solving options.

The first step in any problem-solving process will be to determine accurately the cost of the problem. This is obtained by clearly developing the problem statement and assigning quantitative values to down-time, quality, customer complaints, etc. When this information has been collected only then can the determination be made to advance.

The first phase is to establish knowledge of the current entrained air and foam levels in the system. Key sample points should be selected and with the use of a portable EGT, the entrained air level can be established at each point in the system. Sample points identified can be before and after fan pumps, screens, cleaners, chests, and deculators. A visual audit of surface foam and deposition should also be carried out. This should include noting levels of turbulence in situations where stock and white water flows into chests which give opportunities for the introduction of air into the system.

To identify the appropriate programme and select the best products, two laboratory test methods are useful. The first laboratory procedure measures the defoamer's effect upon surface foam from the perspectives of: immediate knockdown and performance endurance. Figure 4.11 shows a photograph of the laboratory equipment used for the test which re-circulates the process sample while maintaining the actual process temperature of the sample. The sample container is transparent which allows the tester to monitor and measure foam levels as they are reduced

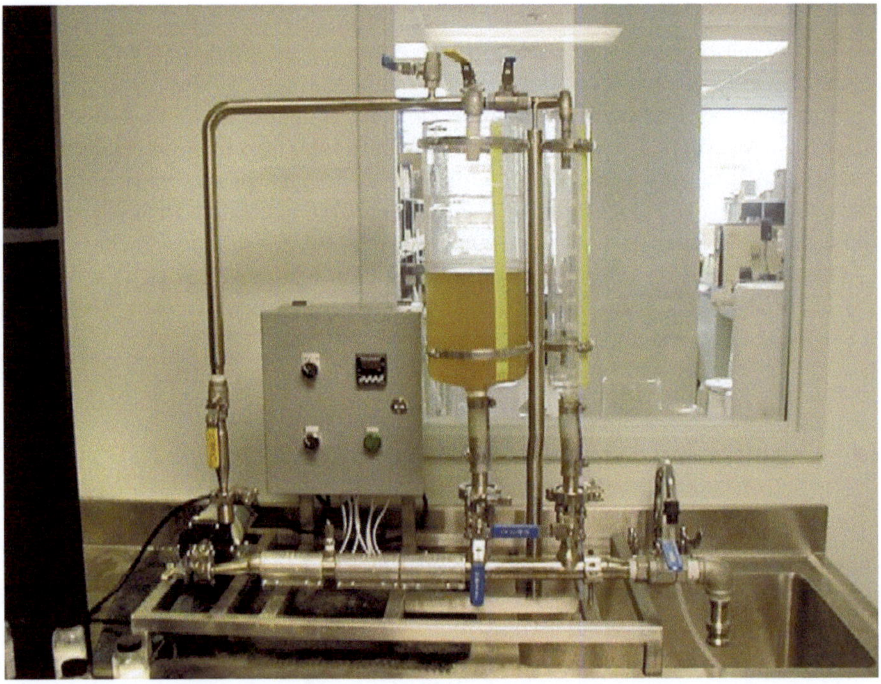

Fig. 4.11 Laboratory foam cell tester used to select a defoamer programme

through the immediate knockdown effect and with time to measure performance endurance of the defoamer.

The second test procedure measures the drainage rate of the brownstock liquor or for paper machine stock. The amount of entrained air in the liquor or stock will negatively impact the drainage curve. And by carrying out testing of different products and feed rates the appropriate product can be selected by comparing performance at equivalent application costs.

4.6.1 Application Strategies

Application and control strategies are evolving from a single application point approach to one of multiple application points for pulp mill and paper mill applications. The single application point approach continues to be applied and can be effective with the advantage that a single product and application point programme is more manageable.

The multiple application point and/or multiple product strategy allow for a two directional approach, i.e. to tackle entrained air and surface foam. The first application point, a base loading of the product and is generally linked to production rates. The second, a trim addition point is where control is carried out. The trim addition on a paper machine application can be linked to an online entrained measurement system, as shown in Figs. 4.9 and 4.10. An example in pulp mill to control defoamer feed rates may be soda carry-over as indicated by liquor conductivity or washing losses.

Defoamers and deaerators, because they are efficient in relatively small quantities, are most useful to assist in managing entrained air and foam problems. However, it must be noted that the incorrect selection or application of these products that are intended to solve one problem can in fact result in other problems in the pulp and papermaking processes. An example is the potential increase foaming tendency, impact on the internal sizing and retention/drainage programmes or the formation of deposits by forming complexes with pulping process by-products.

As explained earlier, it is very important to quantify as carefully as possible the value of the problems caused by foam or entrained air. It makes no sense to spend 200,000 to fix a problem that costs 100,000. In the same way it is crucial to look carefully at all the problems foam is causing. Here are just a few examples that a monetary value can be attached to:

Poor formation – Small bubbles in the stock disrupt the sheet.
Pinholes – Bubbles can cause voids in lightweight paper.
Porosity – Entrained air can make a sheet more porous; this can increase uptake of coating.
Drainage – Entrained air slows movement of water through the sheet.
Increased steam use in drying – Poor formation and poor drainage mean a wetter sheet enters the dryers.

All factors that cause foam in the system should be evaluated. But very often the most efficient way to manage the problem is to use an effective chemical defoamer.

4.6.2 *Troubleshooting*

This section provides information on troubleshooting foam control applications. In a process running under steady-state conditions a significant change in defoamer consumption should alert personnel to investigate and start the troubleshooting process to ascertain the cause for the change in defoamer consumption. The following list provides investigation items that should be incorporated into any effective troubleshooting process:

- Measure entrained air – Use a portable EGT to sample inlet/outlet of key process equipment.
- Confirm defoamer application rates and feed equipment operation.
- Make process audit for visible foam: white-water trays, silo, tanks and sewer.
- Product quality – Is the final paper/board quality affected with holes, formation, deposits or other sheet defects?
- Chest and tank levels – Check changes in turbulence due to flow or mixing.
- Pump condition – Check for worn and/or leaking seals and packing.
- Reverse/forward cleaners and other screens – Ensure seals are intact, inlet/outlet pressure and that the accepts/rejects ratio is within normal operating conditions.
- Wet-end additives – Are the wet-end chemical additives within normal consumption rate ranges?
- Process condition changes – Carry-over from the pulp mill, furnish composition, system pH, temperature and conductivity.
- Process equipment – First, is the equipment in operation, and second, is it operating within the set operating ranges?
- Check for external contamination, e.g. black liquor contamination of condensate used for final stage washing.

4.7 Summary

The impact of air in pulp suspensions is critical to the pulp producer and papermaker from a standpoint of runnability, process cleanliness and quality. Due to the difficult economic challenges facing both pulp producers and papermakers, major capital expenditure to correct design issues or replace major pieces of equipment may not be a viable option. However, the regular maintenance and upkeep of equipment will assist in minimising problems associated with air introduction into the process and foam creation. The appropriate selection of defoamer or deaerator products in combination with application knowledge is one option for papermakers to manage foam and entrained air to bring improvements in the production process. The challenges faced by product development chemists centre on developing a product that combats the pulp and papermaking process which is more susceptible to foam generation; closed loop water circulation; increase in system temperatures; larger and faster paper machines; and gap former technology. Advances in defoamer chemistries, i.e. water-based silicone technology is environment-friendly and can reduce carry-over and increase effectiveness. They have assisted in alleviating resin and deposit problems associated with conventional oil-based defoamer products. For example, silicone defoamers used in the pulp showed significantly less carry-over when compared to oil-based products (Mudaly and Stranges, 2006), leading to a reduced potential of defoamer-related deposition further along the pulping and papermaking processes. More effective defoamers in the pulp mill can often solve serious quality problems in the paper mill (Hoekstra, 2007).

There is no substitute for 'hard work' to survey the process systems and learn the pulp and papermaking process to understand potential incompatibilities and synergies with other wet-end additives and the process conditions, followed by laboratory studies evaluating different defoamer chemistries. Defoamers should be evaluated for their immediate effect to 'knock down' surface foam and achieve longevity in performance, as well as impact entrained air.

The use of online measurement and loop control devices provide 'real-time' information, incorporating a function of control and providing the papermaker with 'ease of mind' that the foam or entrained air is maintained within set limits and under control.

As a final note, when faced with the problems related to entrained air and surface foam the following will assist toward successful evaluation of a selected defoamer product in a pulp or papermaking system: baseline data collection, identification and quantification of avenues that lead to a positive return on investment, and a mutually agreed-to evaluation plan along with implementation phases following the evaluation plan that allow for day-to-day operation changes with contingencies.

Acknowledgements The authors wish to acknowledge the special and invaluable contribution made to this chapter by Mr. Angelo N. Melchiorre, Mr. Patrick D. Blaskowski, Mr. Glenn D. Mudaly and Mr. Ryan Forrest.

References

Garrett, P. R. (1980). Preliminary considerations concerning the stability of a liquid heterogeneity in a plane-parallel liquid film. *Journal of Colloid Interface Science*, *76*(2), 587–590.

Gysling, D. and Loose, D. (January 2004). Clamp-on sonar based entrained air measurement for pulp and paper applications. 90th Annual of Pulp and Paper Technical Association of Canada. Montreal, Quebec Canada. vol. B, 109–111.

Helle, T. (2000). Qualitative and quantitative effects of gas content on papermaking. *Paperi Ja Puu 82(7)*, 457–463.

Helle, T. and Paulapuro, H. (2004). Effect of precipitated gas bubbles in papermaking Appita Journal *57*(6), 444–447.

Helle, T., Meinander, P. O., Nykaneen, R. J., Molander, K. S., and Paulapuro, H. V. (1999). Air removal mill trials using pomp deaerator. *TAPPI Journal*, *82*(6), 146–149.

Hoekstra, P. (2007). Improving Washing Efficiencies in the Kraft Pulp Mill with New Defoamer Technology. TAPPI 2007 Engineering, Pulping & Environmental Conference. Orlando. 22–24.

Matula, J. P. and Kukkamaki, E. (2000). New findings of entrained air and dissolved gases in pm wet end: Mill case study. TAPPI Journa *83(4)*, 22.

Mudaly, G. and Stranges, M. (2006). Bowater Thunder Bay Ontario, kraft mill reduces defoamer usage with new siloxane technology. PAPTAC 92nd Annual Meeting. Montreal: Quebec Canada. 2006 6–10 Feb, vol. B, 231–236.

Pietikainen, T. (1992). On-line air entrainment tests on paper machines. *TAPPI Journal*, *75*(11), 185–191.

Rauch, R., and Burke, T. (July 2002). Optimised deaeration leads to substantial process and quality improvements in paper manufacturing. *TAPPSA Journal*, 22–27.

Schwarz, M., Gmeiner, A., Mannes, W., and Binder, E. (2002). Gear up the wet end. *Pulp & Paper Europe*, *7*(2), 36.

Taber, P. L. (2003). Analyses show entrained air as a contributing factor in sheet barring. *Pulp and Paper*, *7*(9), 46–48.

Wang, G., Pelton, R., Hrymak, A., Shawafaty, N., and Hend, Y. (1999). On the role of hydrophobic particles and surfactants in defoaming. *Langmuir*, 15, 2202–2208.

Watzig, D. and Rauch, R. (2004). Techniques for measuring stock properties. *Industria Della Carta*, *42*(4), 60–72.

Woodworth, M. (1990). Compression of practical methods for air-stock measurements. *TAPPI Journal*, *73*(11), 135–138.

Chapter 5
Sizing in Acid, Neutral and Alkaline Conditions

Dave Johnson

5.1 Introduction

The selection of a sizing agent for any paper or boardmaking application is very often subject to the end requirements of the paper or board. There is a choice of sizing agents available, and currently the main types are:

- Rosin-based sizing agents
- Alkyl ketene dimer (AKD)-based sizing agents
- Alkenyl succinic anhydride (ASA)-based sizing agents

The major requirement for any sizing agent is efficient sizing, and with the cost constraints being increasingly applied to paper- and boardmakers worldwide, this must also include cost-efficient sizing. In recent years, sizing chemical suppliers have come under increasing pressure from paper- and boardmakers to optimise their costs and efficiencies. In the past, tailor-making a sizing product or system for a particular mill or grade of paper was normal, but sizing suppliers have had to reduce their range of products and concentrate on optimising wet-end systems to make a number of standard products work. From this point of view, the selection of sizing agents, retention systems and other chemical additives have become more closely linked. In fact, most size suppliers now offer a range of retention systems and other wet-end chemicals, to give the best benefits for their sizing agents and offer 'packages' in order to optimise the efficiencies of sizing products. However, with the technology that now exists, with online retention and charge monitoring, sizing and retention chemical addition rates can be optimised to give the best, most cost-efficient sizing (and retention).

It was once remarked at a papermaking seminar, when the topic of sizing was discussed after the topic of retention, that the lecturer on retention mentioned sizing only once or twice during the presentation, but the lecturer on sizing mentioned retention almost as much as sizing, indicating the importance of good retention to good sizing.

D. Johnson
Eka Chemicals Ltd., Bridge House,
High Street, Congresbury, Bristorl BS49 5JA, UK
e-mail: Dave.Johnson@akzonobel.com

I. Thorn and C.O. Au (eds.), *Applications of Wet-End Paper Chemistry*,
DOI 10.1007/978-1-4020-6038-0_5, © Springer Science+Business Media B.V. 2009

However, there are many factors to consider when deciding on the sizing product to be used, some of which are:

- The type of paper or board produced
- The pH of the wet-end of the paper machine
- The use of filler and type of filler
- The retention system
- The end use of the paper or board
- Customer requirements
- Cost

For example, it is no use using large amounts of expensive sizing agent for a low-cost product, such as newsprint, unless it is possible to obtain a premium price for the product. Other examples are using rosin size in highly filled paper with calcium carbonate, when the use of AKD/ASA is far more efficient due to the system pH.

Also some types of sizing agent are unsuitable for certain grades because they do not give the final end-use properties, such as liquid packaging board, where AKD is the preferred size for standard grades because it can give lactic acid resistance, and AKD/rosin are the preferred sizes for aseptic grades because AKD gives lactic acid resistance and rosin is required to give hot hydrogen peroxide resistance.

However, the general trend is towards neutral and alkaline papermaking, so currently the predominance is with cationic rosin sizes for neutral and ASA/AKD sizes for alkaline. The difference in the three types can be described by:

- Product type
- Application method

5.2 Product Type

5.2.1 Rosin Sizes

Rosin can be supplied in three basic forms:

- Rosin soap
- Anionic rosin emulsion
- Cationic rosin emulsion

Rosin is a non-reactive product and is retained on the anionic fibre by either attaching itself to a cationic source in the case of rosin soap, or anionic rosin emulsion or by having a cationic surface charge. The main requirement is that it requires a source of aluminium species to form the actual sizing agent, aluminium rosinate. Rosin is normally modified with maleic anhydride or fumaric acid to increase its reactivity with aluminium species and improve its efficiency at higher pH. The method by which this aluminium species is formed and retained in the wet-end of the paper/board machine differs between anionic and cationic rosin sizes.

Rosin soap size is a soluble sodium or potassium salt of rosin and, as such, is highly anionic. Papermaker's alum ($Al_2SO_4.18H_2O$) is added to the stock and at acidic pH (4.0–4.8) is cationic as the Al^{3+} species, so is attracted to the anionic fibres and fines and the rosin soap size. This provides an electrostatic bridge between the highly anionic rosin soap and the fibre/fines. The rosin distribution is normally poor initially, and can be worse in high hardness water due to the neutralisation of fibre/fines charge by more highly active calcium and magnesium ions. The aluminium rosinate is formed and melts, or 'sinters', in the drying section of the paper machine. For rosin soap, the 'sintering' temperature is high because the aluminium rosinate is mainly present as the di-rosinate with a small amount of free rosin. The consequence of this is that more rosin is required to obtain sizing than with rosin emulsions. Anionic rosin emulsions are dispersions of the 'free' rosin with a small amount of dispersing agent, or surfactant, as the sodium or potassium salt. The retention mechanism is similar to that of rosin soap size, but as there is less 'soap' present, less alum is required to form the electrostatic bridge, which allows the system to be used at a higher pH. Fortunately at higher pH (~5.2), alum also exists as a polymeric species, $(Al_8(OH)_{10}(SO_4)_5)^{4+}$, which is more cationic than Al^{3+} and so is more efficient for sizing. Also with rosin emulsions, only a small amount of aluminium rosinate is formed as the mono-rosinate with a large amount of 'free' rosin acid (1). This has the effect of reducing the 'sintering' temperature, allowing the rosin to spread more easily and giving more uniform coverage with less drying required.

Because of the lower levels of aluminium species required, the options for size addition with rosin emulsions are much greater, such that it is possible to add rosin emulsions and aluminium species close together, or premix in a controlled way, at pH up to 7.0. By doing this, the rosin emulsion can be essentially cationic at the point of mixing with the fibre and, if sufficient aluminium species is present, give effective sizing under what is normally considered unrealistic conditions for rosin sizing.

However, under these conditions, the retention of the aluminium species-rosin complex becomes more difficult, so the use of efficient retention systems becomes imperative.

One negative effect of increasing the pH is that for some grades of paper made from unbleached fibre in integrated paper mills, there is a considerable amount of 'natural rosin' present, which at low pH is available for sizing. However, as the pH is increased, and this rosin is not modified to improve performance at higher pH, its efficiency is decreased dramatically and its contribution to sizing decreased to almost nothing. This can give rise to an increased sizing cost on these grades if papermaking at higher pH is required.

This premix of rosin emulsions and aluminium species has led to the development of cationic rosin emulsions for sizing at 'pseudo-neutral' pH (6.0–7.0).

Cationic rosin emulsions come in three forms:

- Rosin emulsion stabilised with cationic starch, or polymer
- Rosin emulsion stabilised with cationic starch, or polymer, with added aluminium species
- Rosin emulsion stabilised with cationic starch, or polymer, with sufficient aluminium species added for sizing and retention

With the first type, an aluminium species will still have to be added for sizing, but considerably less than that required for the appropriate anionic rosin emulsion. However, the addition of the aluminium species needs to be close to, or with, the cationic rosin emulsion to avoid the problems of aluminium chemistry at neutral/alkaline pH.

For the second type, even less aluminium species is required for sizing, and addition point of aluminium species is not as critical as the first type, but as the objective is to use the product at higher pH with less aluminium species, the same addition point strategy for the aluminium species applies.

The third type is essentially a 'one-shot' rosin sizing system for use at 'pseudo-neutral' pH, and in very clean systems has been shown to be efficient. However, in practice, such systems are 'few and far between' and additional aluminium species is often required to maintain good sizing efficiency.

However, the more the aluminium species that is added to these products, normally in the form of liquid alum, or polyaluminium chloride (PAC), the lower the active rosin content, so transportation costs increase, relative to active product. From this point of view, the most effective products are the higher active rosin content, type 1, or type 2 products.

One of the problems seen when sizing with rosin at 'pseudo-neutral' pH is the 'loss of sizing' from the natural rosins in unbleached pulp, especially in integrated mills. This natural rosin can often contribute positively at pH 4.5–5.0, but as it is not 'modified' in any way it loses its effectiveness at higher pH.

These problems will be dealt with in more detail later.

The most critical chemical in rosin sizing is the source of aluminium ions to react with the rosin size, as unless the aluminium species is contained in the product, as in type 3 of the cationic rosin sizes, there will be no sizing. It is possible to obtain sizing with other metal ions, but these are much less efficient, or more costly, than aluminium. As previously stated, the two most commonly used chemicals for this purpose are aluminium sulphate (papermaker's alum) and PAC.

Aluminium chemistry is a highly complex subject, which could easily merit a chapter of its own. However, it has proved possible to use alum at pH values (6.0+), where it was previously not thought to be effective, by the correct application.

Although more expensive, PAC is seen as an alternative to alum, particularly at higher pH. PAC dissociates in solution to give $(Al_{13}O_4(OH)_{24})^{7+}$ and has the advantage of being more cationic than alum, as shown in Fig. 5.1, which should make it more reactive and more effective for rosin retention at the wet-end; it also has the advantage of not putting sulphates into the wet-end which could cause biological problems.

5.2.2 Alkyl Ketene Dimer Sizes (AKD)

AKD sizing has been available for almost 50 years and was the first commercially available 'Synthetic, reactive' sizing agent. AKD is a wax normally derived from a mixture of stearic (C18) and palmitic (C16) fatty acids with a melting range between 42°C and 55°C, dependent on the fatty acid blend. However, there are

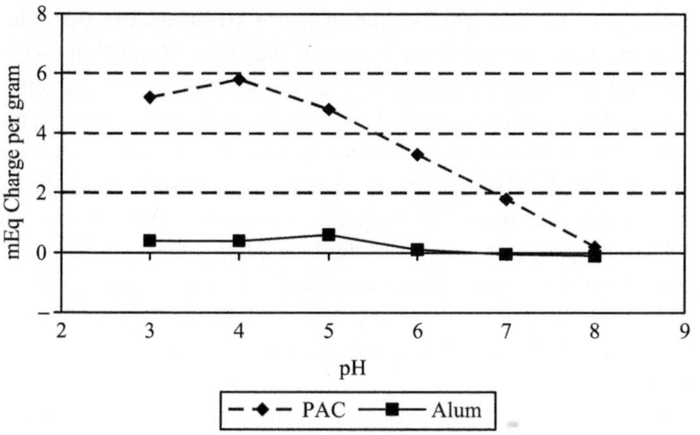

Fig. 5.1 Effect of pH on charge of alum and PAC

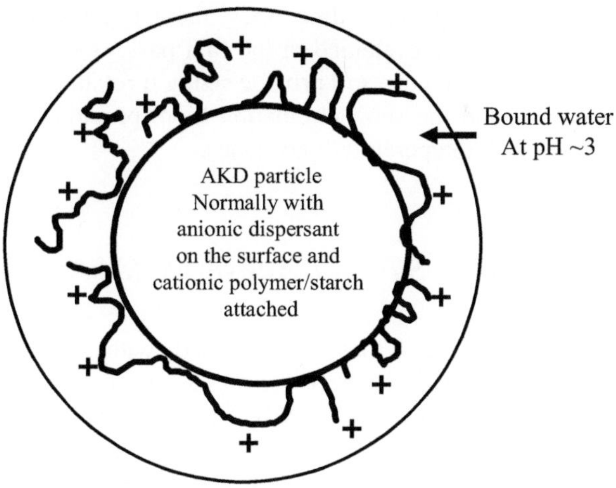

Fig. 5.2 The AKD particle

alternative types based on unsaturated (AKD) and branched fatty acids, but these are in a very small number of cases, for specific applications. The product develops sizing by breaking of the ketene ring reacting with the hydroxyl (OH^-) group of cellulose fibre, it is most efficient in a neutral, or preferably mildly alkaline (7.5–9.0), pH environment. The product is normally supplied to the customer in the form of a dispersion. The dispersion is made normally with cationic starch and/or cationic polymer. The dispersion is active and stable for up to 3 months at 20–25°C. This activity/stability can be greater at lower temperatures, or lesser at elevated temperature. The ideal storage temperature is below 20°C. The degree of sizing achieved with AKD at the wet-end is related to the retention of the AKD and the drying of the paper, or board, produced. AKD is normally slow reacting with the fibre and the dispersion incorporates a bound layer of water at low pH to minimise hydrolysis as illustrated in Fig. 5.2.

It is essential to remove this water in order to break the particle and allow the AKD wax to melt, spread and react with the fibre. Retention of the AKD can be improved by the use of additional cationic starch, or cationic polymers. The starch, or polymer, may be added separately, or included in the emulsion with the AKD wax. The role of this additional starch, or polymer, can be to stabilise the AKD dispersion, but is often more of use as an anionic trash collector (ATC) as it is more cationic than the AKD itself and will neutralise the anionic trash, allowing the AKD to be preferentially adsorbed on fibres, fines and fillers, increasing the retention of the AKD wax. If incorporated into the dispersion for stability, the removal of the polymer to act as an ATC will destabilize the dispersion, allowing it to break more quickly in the drying section and thus increase the reaction of the AKD with the fibre.

The retention of the AKD, the breaking of the dispersion, and subsequent sizing development, will be influenced by the amount and type of cationic polymer/starch used, either in the dispersion, or added separately to the stock, and the amount of anionic trash present in the furnish. Good retention of the AKD is essential for efficient sizing, but unlike ASA, there is considerable evidence that the AKD dispersion, if not retained in the first pass, is still in an active form whilst in the white water, and can still provide sizing if retained on second, third and subsequent passes round the short circulation loop, provided that the conditions are not too harsh (high temperature/high shear).

5.2.3 Alkyl Succinic Anhydride (ASA)

ASA is like AKD in that it reacts directly with cellulose fibre at neutral pH. However, it is based on derivatives of the oil industry and the reactive group is the anhydride group. This is far more reactive than the ketene group of AKD. ASA is generally made from C16 and C18 olefins and is a liquid (oil). It differs from AKD in that it is approximately half the chain length of AKD and the reactive anhydride group can be at any point on the chain, whereas the ketene group in AKD is always central because the product is a dimer. Because the chain length is shorter than that of AKD, it is less efficient as a sizing agent for the same amount of reacted product. However, the reaction between the cellulose and the ASA is more rapid than with AKD, but also the amount reacted is greater, which gives an advantage of ASA over AKD. This means that in terms of efficiency, the products are fairly similar, although in ideal conditions for ASA sizing, ASA may be more efficient, but in conditions ideal for AKD sizing, AKD will be more efficient. ASA reaction is normally very fast, however, this is by no means guaranteed, particularly where lower drying conditions are prevalent (e.g. waste-based paper and board). The ASA particle behaves in the same way as the AKD particle, in that it must break in the drying section before it can spread and react. However, under poor drying conditions, or under high moisture conditions, this may not always take place on the paper/board machine and may lead to a curing effect in reels. Certainly, sizing development is much more rapid than using

AKD on an equivalent furnish formulation. This fast reactivity of ASA can be its strength and its weakness. Because of the increased reactivity, optimum sizing development can be achieved between pH 6.0 and 8.0, at a lower starting pH than AKD, but not as high a pH as AKD, as the hydrolysis with ASA can become more rapid at high pH. The reactivity with water is just as rapid as with the cellulose fibre and the speed of reaction also increases at higher temperatures, so good retention of the ASA dispersion is of even more importance. This was a problem in the early applications of ASA. The consequence of the ASA reactivity with water is to place an enormous importance on good retention levels in the sheet. As with all cationic sizes, the size is attracted to the particles in the wet-end system with the highest charge density, the fines and filler. High retention of the fines and filler and consequently the ASA size in the sheet prevents a major problem. If ASA is not retained, it will rapidly convert to hydrolysate of the size. This hydrolysate is reactive towards calcium and magnesium and if these ions are present, which they are in almost all alkaline papermaking systems, they will form a 'half acid/half metal salt', which can give rise to very sticky deposits. However, in recent times, the use of more efficient micro polymer and even more importantly micro/nano particle retention systems, with improved application techniques, has reduced this problem.

One advantage of ASA compared to AKD is that the reactive group, the anhydride ring, can be situated at any point on the chain. Depending on the position on the chain, the product may be more, or less, reactive and more, or less, hydrophobic. In general, the most reactive tend to be the least hydrophobic, so the structure of ASA is generally a compromise between the two properties. Unfortunately, as ASA is a by product of the oil industry, it can be subject to changes in the world oil price. Most ASA is made by the oil companies, not the paper chemical suppliers (with a few exceptions), so there is very little choice of reactive group position and mills have to 'take what they can get'. However, the major ASA suppliers to the paper and board industry, including the ones that produce their own ASA from the Olefin source, have used their knowledge to ask for more specific Olefin/ASA chain lengths, and 'higher purity' ASA sizes (95–100% C18) are now becoming more prevalent, compared to previously used 35–65%:65–35% C16/C18 mixtures.

ASA size is normally added to a paper machine as a dispersion made on the mill site, normally using a small portion of the mill cationic starch, although some paper-makers do not use cationic starch, or the cationic starch is not suitable, in this case a cationic polymer can be used. The dispersion is produced through an automatic dispersion unit, using various different techniques, depending on the supplier/equipment type and these are normally supplied by the sizing supplier.

5.2.4 Other Sizes

Other sizes include:

- At the wet-end
  Paraffin wax emulsions precipitated with alum

- At the size press/film press
  Polyurethanes
  Styrene maleic anhydride polymers
  Styrene acrylate polymer emulsions

The use of paraffin wax emulsions is very rare, but they have been used for highly water-resistant grades of board. The anionic wax dispersion is precipitated onto the fibre with alum. It generally relies on the very high self-retention of high basis weight board grades for efficiency and, because it is not always correctly orientated to the fibre surface, is much less efficient than conventional sizes.

The sizes added at the size press/film press are commonly used on modern paper machines to impart a particular property to the paper which cannot be achieved by wet-end size addition alone. These properties include:

- Reduced sizing cost (size press/film press size is 100% retained)
- Better toner adhesion of printing inks
- Better colour definition with colour/ink jet printing papers
- Very hard sizing specifications
- Friction coefficient changes

However, as this book is concerned with the practical application of wet-end chemicals, these should be covered in other articles, but it is sufficient to say that good wet-end size application is required to allow these surface applied chemicals to work in the correct way.

5.3 Application

For the application of sizing agents the following should be considered:

1. Paper machine type, speed, paper grades:
As an example, machines that have an MG cylinder have had problems converting to AKD/ASA sizing, due to the loss of adhesion to the MG cylinder, because of the addition of 'waxy' AKD, and to a slightly lesser extent, 'oily' ASA. They tend to still use rosin sizing and attempts to raise the pH of such systems have been using the 'Pseudo-neutral' rosin sizing route. Also, in general, higher speed machines tend to put more 'pressure' on the drying and require faster 'curing', so the tendency is to use rosin size at acid pH and ASA sizing at neutral/alkaline pH which are more 'instant'.

2. Type of fibre:
The choice of sizing agent and application technique can vary dramatically depending on the fibre. Normally, recycled fibres require more sizing agent than virgin fibre, so the choice of sizing agent is critical for economic sizing with recycled fibre. Another example is unbleached virgin fibre, which often contains residual rosin, that can reduce the requirement for additional rosin sizing agent to very low

amounts when acid sizing with alum. This is why many grades of packaging made with unbleached virgin fibre are still made under acid sizing conditions

3. Type and amount of filler:
The filler type is probably the most important factor in the selection of a size. The use of calcium carbonate ($CaCO_3$) normally means that sizing will take place at neutral to alkaline pH, as $CaCO_3$ normally buffers the pH to 7.7. This would mean that AKD size, or ASA size, would be the preferred sizing agent. However, depending on the form of the $CaCO_3$, whether it is ground $CaCO_3$, chalk, or precipitated calcium carbonate (PCC), the amount added and the grade of paper, the choice of AKD, or ASA currently tends to be towards ASA. PCC-filled paper also tends to be run at a slightly higher pH (8.0–8.5) than ground calcium carbonate (GCC) filled paper, due to residual lime ($Ca(OH)_2$), present as a by-product of the production process. However, both ASA and AKD have been used successfully on highly filled printing and writing paper, especially in combination with effective surface treatment.

Clay, has traditionally been used with rosin sizing systems, although there is no reason that AKD, or ASA, should not be used providing the correct system pH can be maintained.

TiO_2 and/or calcined clay are used in paper grades where high opacity is required. They can be used with any sizing agent and can be difficult to retain because of their small particle size and high surface area. They also have a high anionic charge and this means that they are attractive to cationic sizes. This means that a large percentage of size can be adsorbed onto the surface of the filler, which in combination with the lower retention mean that size additions tend to be high when these types of filler are used.

4. Incoming water – time of year:
The quality of the incoming water can adversely affect the retention of cationic sizes, if there are high levels of anions present which adsorb, the cationic size, or also if there are metal ions present which can neutralise potential sites for size retention on fibre.

5. Refining degree:
Refining can create higher levels of fines in the system and these fines have a higher surface area and higher anionic charge, which means that, unless they are agglomerated by other cationic additives, more sizing agent is required to cover the surface area.

6. Stock additives, interference from other chemicals:
Other stock additives can have both a positive and negative affect on sizing. Negatively, anionic additives can neutralise cationic sizing agents and cationic additives can be preferentially adsorbed, positively some anionic additives can provide a site for retention of cationic sizing agents and cationic additives can neutralise anionic trash, giving better retention of cationic sizes on fibres.

7. Stock temperature:
High stock temperatures can have a negative effect on all the major size types, causing agglomeration of rosin types and hydrolysis of reactive sizes.

8. Total acidity/alkalinity - pH:
The pH of the system and the acidity/alkalinity tend to decide the type of sizing agent required. In general rosin size tends to be favoured in acidic conditions and ASA, or AKD, sizing in alkaline conditions.

9. Sequence and timing:
The addition point and time of exposure to conditions detrimental to sizing, or sufficient adsorption/retention time and sequence relative to other additives can affect sizing both negative and positively.

10. Retention/Drainage:
Retention of sizing agents is probably the single most important factor in sizing efficiency, as much of the size is initially attracted to fines and filler and retention of these is paramount to good sizing. Drainage, when associated with good retention, will enhance sizing, as it reduces the amount of water to be removed from the size particle. However, excessive drainage can result in loss of fines and filler, and loss of size associated with these.

11. Formation:
Good formation is essential for good sizing, as otherwise, there will be 'holes' in the paper/board, which allow liquids/inks to penetrate.

12. Press moisture:
Lower press moisture will allow more heat transfer in the drying section enabling faster breaking and spreading of the size and faster reaction for ASA/AKD.

13. Drying conditions:
The ideal drying conditions for ASA and rosin sizing differ from those of AKD, whereas for ASA and rosin there should be mild drying in the early drying section, AKD benefits from more extreme temperatures.

14. Sizing specification required:
As has been previously mentioned, some sizing agents cannot give the required specification of the paper or board, and may be eliminated immediately for that reason alone. For example, plaster (gypsum) liner board has a sizing specification that can only be achieved by a product resistant to boiling (hot) water. AKD is unsuitable as it is not always fully reacted with the fibre and can be removed by the boiling/hot water. The waxy nature of the AKD can also prevent the plaster sticking to the board when it is applied. So, plaster liner board is normally sized using rosin and alum, or ASA.

15. Surface treatment/coating:
The addition of surface treatment, whilst generally complimentary to wet-end sizing can be negative if the sizing is not properly 'developed' prior to the surface treatment. An example of this is on-machine coating where pre-coater sizing is essential to achieve good coating holdout. If sizing is not achieved before the coater, then there is a preferential water pickup compared to the coating. This causes the viscosity of the coating to increase, giving wrong coat weight and poor

runnability. Also the coating mineral will return to the wet-end of the paper machine via the broke. Depending on the type and surface area they can influence the size selection, as in fillers.

16. Calendering:
As in surface application, calendering can have a positive influence on sizing, if properly developed prior to calendering. However, if the sizing is not properly developed, this can be a negative influence, especially with poorly dispersed rosin size or poorly reacted/fixed AKD size.

5.3.1 Application of Rosin Size

Rosin is supplied as a dispersion, or soap, and the application is normally easy. It has a fairly long shelf life, which can be up to 6 months, so storage considerations are not an issue, although extremes of temperature and storage in unsuitable containers should be avoided, but these are normally recommended by the size supplier. All that is required is a suitable pump and the correct choice of addition point. Normally, it is preferred to add the rosin size at a point of good mixing, such as the inlet of a pump, or a point of high turbulence. If this is not possible then dilution should be considered. The quality of the water used for dilution is important and fresh process water should always be used. The dilution should preferably be just before the addition point, as high alkalinity in some 'fresh' water can be detrimental to the performance of rosin size with prolonged contact. For all anionic rosin emulsion sizes and cationic rosin emulsion sizes, high incoming stock pH (8.0+) should be avoided as this can cause 'soaping' of the free rosin and formation of free rosin metal salts (Na^+/Ca^{2+}) which will lead to poor sizing. However, cationic rosin emulsions should not be added close to other highly cationic additives (apart from aluminium species used for sizing), whereas anionic rosin emulsions and anionic rosin soaps can benefit from addition close to cationic species.

One important factor for rosin sizing at 'pseudo-neutral' pH is the point of addition of the rosin size, be it anionic, or cationic. In general, the higher the headbox pH, and the higher the stock temperature, the later the addition point of the rosin/aluminium species should be, in order to reduce the exposure of the rosin/aluminium complex to unfavourable conditions for rosin sizing. Of course, this is very much dependent on conditions on the paper/board machine and very late addition into systems with high white-water fines content should also be avoided. However, the one of the most important factors is good retention of the size and aluminium species, especially at higher pH (6.0+). If the rosin particle is stable in the wet-end of the paper machine and conditions, such as high temperature and pH, are not unfavourable, the product can be retained on subsequent passes and react with additional aluminium species to give sizing. However, if it has already combined with aluminium species, then it will already be subject to the conditions that are favourable/unfavourable for aluminium chemistry and at higher pH (6.0+) can give rise to deposit problems.

5.3.2 Application of AKD Size

As with rosin size, AKD size is supplied as a dispersion and the application is normally easy. It has a reasonable shelf life, which can be up to 3 months, but storage considerations can be an issue and high storage temperatures, above 30–35°C should be avoided. However, provided the product is stored in a cool place, it can and has been successfully used in many countries with high ambient temperatures. At these temperatures, the shelf life is reduced and once exposed to these temperatures it should be used within one month. Again, as with rosin size, all that is required is a suitable pump and the correct choice of addition point. It is preferred to add the AKD size at a point of good mixing, such as the inlet of a pump, or a point of high turbulence. If this is not possible then dilution should be considered. The quality of the water used for dilution is important and fresh process water should always be used. The dilution again should preferably be just before the addition point, as anionic contaminants in some 'fresh' water can destabilise cationic AKD dispersions with prolonged contact. The application of AKD sizes is less of a problem than that of rosin size as there is no problem of aluminium chemistry to overcome. However, the most critical application factors are similar to cationic rosin sizes in terms of temperature, as stock temperatures in excess of 50–60°C are as detrimental to AKD sizing at neutral to alkaline pH as to rosin sizing. One of the most important factors is the pH of the system. AKD is most efficient at alkaline pH and a pH of 7.5–8.5 is the preferred operating range. The efficiency of AKD is reduced below pH 7.0 and the hydrolysis is increased above pH 8.5, if not retained, so as with rosin size, the retention of the AKD is important. In the same way as rosin size, as it is a reactive sizing agent, the fact that the reactivity is slow means that, although it is preferred to retain the AKD in the first pass, provided the AKD particle is stable in the wet-end of the paper machine and conditions, such as high temperature, are not unfavourable, the AKD can be retained on second, or third passes, and still be an efficient sizing agent.

5.3.3 Application of ASA Size

Unlike rosin and AKD sizes, because of its reactive nature, ASA cannot be supplied as a 'ready to add' product. It has to be made into a dispersion at the paper machine and added within a short time, normally less than 30 min. This means that it has to be mixed with a cationic component and emulsified with equipment which is supplied by the size supplier. This was an expensive and problematic area, with the costs being added to the price of the ASA itself. But in the past 10 years, technology has improved so that the application costs have been reduced to only slightly more than a sophisticated pumping system and the application technology has improved the reliability/efficiency of the product. This has meant that ASA has changed from a product of necessity to a product of choice. The ASA is supplied as an oil and is stable for more than 1 year, if not exposed to moisture. As stated, it needs to be made into a dispersion with a cationic component. This product is usually the mill cationic starch, but if the mill does not use cationic starch, then the size supplier will supply a suitable cationic polymer. Normally the amount of starch, or polymer, required for this is between 30% and 100% of the required ASA amount, representing a very small portion of the normal amount of cationic starch used on a paper, or board machine. However, not all cationic starches are suitable, those such as high pH types (pH 6.0+) are unsuitable, and it is possible that changes to the starch, or change of starch type, may be required for efficient ASA sizing. In addition, a portion of the wet-end starch is normally added with the ASA dispersion to aid the retention of the ASA, so called 'secondary starch'. This would normally dictate that a minimum of around 3–4 kg/t of cationic starch is required for ASA sizing. It must be remembered that the starch used for emulsification/dispersion (but not the secondary starch) is sheared and is no longer available for strength/retention and cannot be considered as part of the starch added for this purpose. Another consideration is the temperature of the starch. If it is too high, i.e. 55–60°C +, then it could cause hydrolysis of the highly reactive ASA during dispersion. If this is the case, then a starch cooler is required, but this can normally be supplied as part of the ASA dispersion equipment, or emulsification control unit (ECU). The product is also diluted after the ECU to aid the cooling and reduce hydrolysis. Some applications use aluminium species in this dilution water to neutralise any hydrolysed ASA made during the dispersion, as aluminium has the effect of neutralising the ASA acid and forming a non-sticky salt. The ECU also acts as a pump and can control the addition of the ASA dispersion, so an additional pumping system is often not required, unless multiple ply application is required and a 'break tank' is used. However, it is often possible to use a flow 'splitter' to avoid this. Normally, it is preferred to add the ASA size as late as possible into thin stock at a point of good mixing, such as the inlet of a pump, or a point of high turbulence, where there is no recirculation, but before the retention system. It has been shown that it is also possible to add ASA to thick stock at the first fan pump, or machine chest pump, provided there is no recirculation back to the thick stock system (level box/basis weight control recirculation loop).

With ASA, good first pass retention of ASA is paramount to good sizing efficiency, as any ASA not retained in the first pass can be easily hydrolysed in the white-water system. This is because the ASA dispersion is not as stable in the wet-end as that of rosin, or AKD, as these products are designed to be stable for a period of months. This hydrolysis product can react with calcium and/or magnesium to form sticky salts, which can deposit and cause many problems. If aluminium is not added to the ASA dispersion, then it should be added to the white-water system to react with any unretained, hydrolysed, ASA to prevent the formation of these salts, by formation of the less/non-sticky aluminium salt. There are applications of ASA that do not use aluminium species, where first pass ASA retention is optimised, but these tend to be in the minority.

5.3.4 Other Considerations

There are many possibilities of product and application, and the aim is to arrive at one product and application which can be utilised on machine without runnability problems and which gives the correct sizing. The basic questions about the operation tend to eliminate one, or both, of the other products.

This initial analysis will then enable investigation of various application techniques/sequences and this is best carried out with the following apparatus/equipment:

- Hand sheet making apparatus, dynamic sheet former or pilot paper machine for sizing and other paper properties
- Britt Jar or Dynamic Drainage Jar to evaluate the retention drainage with each application
- Particle charge analyser or Zeta Fibre Potential meter to determine the effect on the charge demand and adsorption of the sizing and other chemicals
- pH meter to ensure the correct pH of application

However, the main objective is to transfer the laboratory simulation to the paper machine. This needs a good deal of planning and preparation, and should include:

- A written trial plan from the supplier to the mill technical and production management with agreed objectives
- Product data sheets and Health and Safety data sheets and any product-handling considerations
- Paper properties required
- Other chemical addition points/interactions
- Agreed addition points

Now that the correct product for the machine has been established, the next section will examine what can go wrong and how to put it right.

5.4 Troubleshooting Sizing Problems at the Wet-End

Most sizing problems can be related to one main problem; the lack of the sizing agent, be it rosin, AKD, ASA, polymeric, or any one of the other synthetic sizes available in the paper or board.

The reasons for this occurring are manyfold and this section will endeavour to reveal the reasons and solutions to the problems that occur. However, there are other problems that can occur, which may not be due to size retention – these will be dealt with at the end of this section. The problems can be summarised into five sections:

1. Lack of sizing
2. Slow cure
3. Loss of sizing, or fugitivity
4. Deposits
5. Other problems

5.4.1 Lack of Sizing

The first things to check when sizing does not occur are the obvious ones. Is the size still being added to the system? Is the pump still working? Are there blockages in the lines or filters? This may seem simple, but it is surprising how many times this has happened on paper and board machines. Everybody assumes that mechanical/electrical systems are infallible, and we all know they are not. Even computers have to be programmed by human beings!

However, assuming all of these items have been checked and everything is working correctly, the next item to check is the size level.

5.4.1.1 Size Level

Is the correct level of size being added to obtain the sizing degree required? The major factors in this are the type of stock coming to the machine, the freeness of the stock and the filler types.

If the stock is highly refined then a higher proportion of fines are present. The surface area of fines is much greater than that of fibre, typically 6–8 m^2/g for a mixed hardwood/softwood blend, compared to 1–1.5 m^2/g for fibres. Obviously the size demand, to cover the surface of the fines, is going to be much greater.

For example, if the fines content of paper or board is increased from 10% to 20%, the size requirement will increase by 26%.

Filler types and levels also have a significant effect on size requirement. Some typical surface areas of fillers are:

GCC: 2–4 m^2/g
Filler clay: 8–10 m^2/g
PCC: 8–12 m^2/g
Talc: 14 m^2/g
Coating clay (no. 2): 12 m^2/g
Rutile TiO_2: 10 m^2/g
Calcined clay: 18 m^2/g
Premium coating clay: 22 m^2/g

As the surface area increases the size requirement increases accordingly.

For example, a mill using precipitated calcium carbonate increased the ash content of the paper from 10% to 20% in several increments. The resulting size increase was 23% to obtain the same sizing level.

A second example concerns the use of coated broke. In this case, the coated broke from an off-machine coater, using premium grade coating clay, was increased from 10% to 40%. This required an increase in size of 11% to maintain the same sizing degree.

5.4.1.2 Lack of Sizing - Rosin Size

For rosin sizing with traditional soap or dispersed anionic sizes, much depends on the alum or aluminium species used. Theoretically, the amount of Al^{3+} ion required to neutralise and precipitate one part of rosin is 0.3. However, other factors control the level of Al^{3+} required, and these have a bearing on lack of sizing with soap or dispersed rosin sizes. These factors are:

- pH
- Alkalinity
- Anionic trash (soluble anionic trash)
- Stock temperature
- Drying temperature

pH

Control of pH, especially when using alum is of great importance as the behaviour of alum changes with pH (see Fig. 5.3). If the pH increases to 6.0, the charge on the alum decreases towards zero. It is important to add the size and the alum as late as possible, in order to decrease the contact time of the alum with the stock. This should ensure that the alum remains in its Al^{3+} cationic form for a long enough time to 'fix' the rosin size to the fibre.

Below pH 5.0, dispersed rosin sizes are less efficient than soap sizes, as only a small (5%) proportion of the rosin is in the soap form and available to react with Al^{3+}.

The mechanism is formation of a polymeric cationic aluminium species on the fibre and fines, followed by attraction of the dispersed rosin. As the polymeric species is only completely formed at pH 5.2–5.5, and at its most cationic, the efficiency of dispersed rosin sizes is reduced below pH 5.2.

Conversely, for soap sizes, as they are dependent on neutralisation of the rosin soap by cationic aluminium species at pH 3.0–5.0, the neutralisation is not efficient, so rosin soaps lose efficiency above pH 5.2 as this polymeric species is formed.

Control of pH is thus very important with rosin sizes, dependent upon the type of size used.

At higher pH values, above 6.0 with dispersed rosin sizes, in order to avoid the hostile environment for alum, a pre-mix system is favoured, which ensures

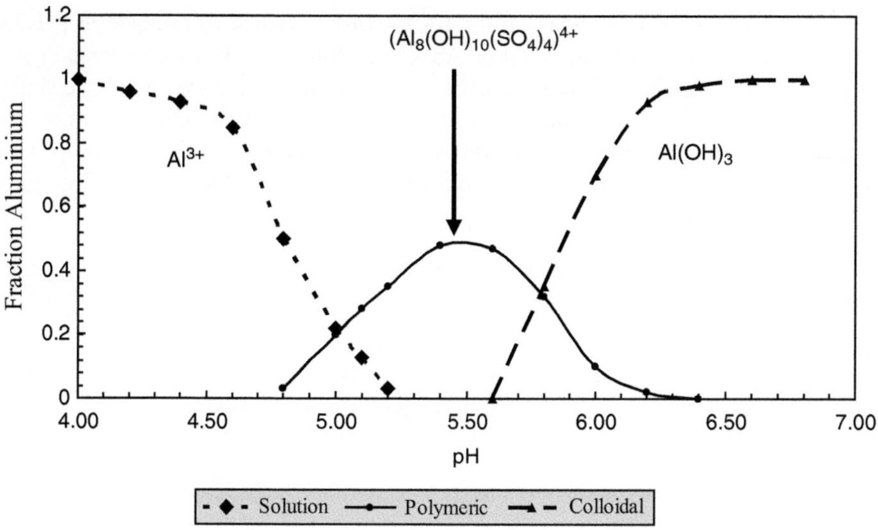

Fig. 5.3 Effect of pH on alum

that some (15–20%) aluminium rosinate precipitate is formed of the correct particle size when mixed with the stock, which then, with further mixing, allows the alum to polymerise and attract the remaining dispersed rosin. This is not the ideal system, but is far more efficient than separate additions. The preferred addition point is late in the system.

An alternative at high pH is to use a compound which maintains its cationicity above pH 6.0 and contains polymeric aluminium species, such as PAC (see Fig. 5.1). As the pH is increased, this can be used in preference to alum and, when used in a pre-mix system with dispersed rosin size, will give more efficient sizing, especially when added late to the wet-end system.

However, the most suitable system is using CRS, as previously described, as most already contain some aluminium species for sizing and cationic polymer for retention, thus eliminating the need to add high levels of additional alum at high pH. These are even more applicable to the pre-mix technique.

Examples of mills experiencing problems with pH are as follows:

1. Mill A, a Kraft liner board manufacturer, was using soap size with alum to size its liner board at pH 3.7–3.9. The mill decided that they were experiencing too much corrosion at this pH and started to raise the pH towards 5.5–6.0. They suffered an immediate decrease in size efficiency with a doubling of size usage. To remedy this, the size supplier changed the size to a dispersed rosin size, the result being a much improved sizing efficiency at the higher pH because of the increased efficiency of dispersed size compared to soap size at this pH.

2. Mill B, producing liner board from 100% waste, wished to produce at pH 7.0 but did not want to use an AKD or ASA size. Initial work with dispersed rosin size and alum at this pH using conventional addition sequence gave little sizing effect. However, to avoid the hostile environment for alum, and to increase the cationicity, they changed to a co-addition of CRS and PAC at the fan pump. This gave them the required sizing without the associated problems of adding alum separately at high pH.

So the questions for lack of sizing when using rosin size are:

- Check pH - is it appropriate?
- Check size - is it the right type for the pH?
- Check addition point - is it late enough if the pH is high?
- Check aluminium source - is it the right type for high pH?

Alkalinity

The presence of alkalinity for rosin size is normally an effect seen with increasing pH, as normally alkalinity increases with increasing pH.

The alkalinity caused by Ca^{2+}, Mg^{2+} and Na^{+} salts give rise to salts of rosin, as they are much more reactive than Al^{3+} in alum. However, Ca^{2+}, Mg^{2+} and Na^{+} do not form polymeric molecules, as alum does at pH 5.5, and are much less cationic than Al^{3+} or its polymeric forms, and are therefore less easy to retain.

The particles formed tend to be smaller than the aluminium flocs, but grow to macro particle sizes. It would be possible to carry out sizing with these salts if retained, but they have very high melting points, so unless the dryer temperatures are 150°C (which is extremely unlikely) the rosin will not be able to spread adequately to give a sizing effect.

Two examples can be used to illustrate how not to deal with alkalinity and how to deal with it properly.

1. Mill A was producing lightweight fine paper from virgin fibre furnish using clay and TiO_2 as filler. At certain times of the year the alkalinity of the incoming water increased dramatically, specifically during times of low rainfall, when it was forced to use a different water source. Using a dispersed rosin size, it increased the addition of alum, to cope with the increased alkalinity from the incoming water. It then suffered from deposits of basic alum, from the reaction of the alum with the $(OH)^{-}$ produced from the alkaline components of the water.
2. Mill B was producing copier grade paper from white waste, virgin fibre and clay, using a dispersed rosin size. It was unable to control the quality of the waste with regard to the amount of chalk and this gave them problems. They sensibly refrained from adding acid or large quantities of alum to neutralise the chalk and sought another solution. To overcome the problem they decided to change to a CRS with alum, at a late addition point, the secondary fan pump. This effectively avoided the alkalinity from the white waste and the machine ran at an efficient size level.

Mill A has since changed to treatment of the incoming water with sulphuric acid controlling at pH 5.5 and has effectively eliminated the variability and deposits caused. Mill B has changed to PAC in its pre-mix system, and has now changed to chalk filler in place of clay achieving much higher brightness paper for the same ash level with similar sizing efficiencies.

In the case of high alkalinity, the solutions to problems are similar to that for high pH, but initially you have to:

- Identify the source of alkalinity.
- Establish if it can be neutralised effectively.
- If it can be neutralised, select the correct chemical for neutralisation, e.g. H_2S0_4.
- If it cannot be neutralised, select the correct sizing regime, i.e. dispersed size, co-addition of alum/PAC, pre-mix of dispersed size with alum/PAC or use of CRS with alum/PAC.

Anionic trash (and soluble anionic trash)

Anionic trash is a problem for all types of sizing, resulting in lack of sizing, due to problems with the retention of these fine particulate materials. With the increasing re-use of waste in paper and boardmaking, these fine particulate materials can be anything from pulp fines to filler and carbon black particles from inks. All of these tend to have a high surface area and tend to be negatively charged.

In integrated mills and with ground wood pulps, this anionic trash takes a different form – that of soluble lignin, anionic contaminants and bleach residues brought into the system from white-water recycling. For rosin sizing, as the pH rises, the charge on this 'anionic trash' increases and it becomes increasingly negative.

As the whole of the action of rosin size depends on precipitation using alum or an aluminium species such as PAC, the preferential consumption of Al^{3+} or polymeric aluminium by this anionic trash means a higher level of alum or aluminium species will be required, initially to neutralise the 'anionic trash' and then to precipitate the rosin. If sufficient Al^{3+} is not added to carry out this, then rosin size will be lost into the white water and consequently a lack of sizing will result.

Some of the lignin present in poorly washed pulp becomes more surface-active as the pH rises and obviously this will interfere with the sizing effect.

The solution to the problem of soluble trash is fairly simple, especially with an integrated mill. Careful control of the washing of the pulp will reduce the soluble trash to reasonable levels and obviously reduce the problem. For non-integrated mills, control of pulp quality is important, but pulp pre-treatment with cationic polymer or ATC is often employed as a pulper additive.

For the anionic trash, the best way of reducing the size requirement of these materials is to reduce the surface area. This can be done by the use of cationic polymers or ATCs. Alum is not as efficient a material as the cationic polymers and ATCs, as the charge is quickly neutralised, especially with alkaline waste pulp. However, reduction of surface area by pre-flocculation is the best way of dealing with this anionic trash. It has the added benefit of reducing the anionic charge, and thus the Al^{3+} demand.

If pre-treatment is not possible, then the only solution is to remove the rosin/alum reaction as far from the anionic trash as possible, i.e. pre-mix sizing or CRS, and reduce the contact time by late addition of size and alum. This, however, will not cure the problem completely, as the high surface area effect will still be present.

There are some components which may be regarded as anionic trash that are beneficial to rosin sizing such as saccharinic acids from alkaline pulping and hemicelluloses from unbleached pulps. These can form complexes with Al^{3+} ions and produce rosin size precipitates of lower melting or sintering points. This promotes more efficient sizing.

An example of problems occurring from anionic trash, of which there are many, is a board mill using a mixed waste furnish. In this mill, waste was not well sorted and was loaded into a pulper in proportions depending on whether it was KLS or white. On a certain grade of board it was found to be almost impossible to obtain the sizing required. A threefold increase in rosin size and a twofold increase in alum were needed to obtain the sizing degree required.

Fortunately, at the time, the size supplier was carrying out a wet-end chemistry/charge survey, and identified a large increase in anionic charge from the pulper. The individual components of the waste were examined and it was found that a proportion of the 'white waste' was of an 'NCR type' with high carbon black content. A pulper pre-treatment with cationic polymer was prescribed by the size supplier. However, the mill decided to eliminate this component and the situation rapidly returned to normal.

So to deal with anionic trash:

- Establish the source of anionic trash.
- Is it soluble from pulp?
- Is it from waste?
- Treat with polymer or ATC.
- Move the size/alum addition point forward.
- Use a premix system.

Stock temperature

Stock temperatures vary greatly from mill to mill. High stock temperatures result from steam injection to improve drainage and aid drying, or aid broke pulping, or as a result of system closure. It can also be caused by seasonal variations in incoming water, or just the temperature of the incoming water in hot climates.

High temperatures are not favourable for rosin sizing for several reasons. Alum forms complex polymers with cationic charge as previously stated. The charge on these polymeric species is reduced as the temperature increases, due to dehydration of this polymer. Also, if the rosin size/alum floc is formed already, the floc is also increased in size, resulting in poor size distribution and putting greater strain on the

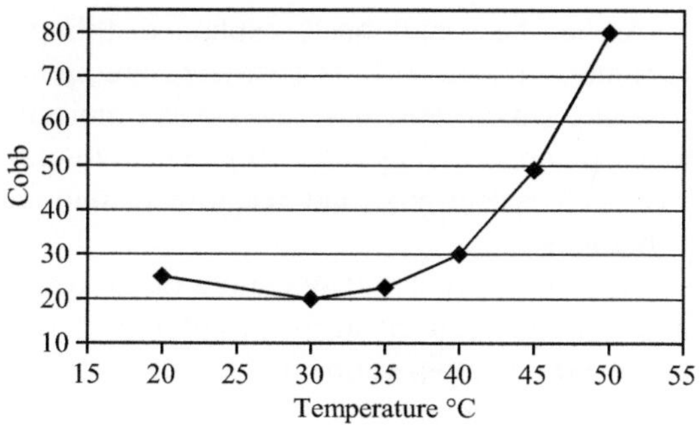

Fig. 5.4 Effect of stock temperature on rosin sizing

drying section, to redistribute the size. The loss of sizing with increasing stock temperature is illustrated in Fig. 5.4. An example of a mill that experienced sizing problems with hot stock was of a liner board machine with a stock temperature of 40°C using 100% waste. The mill closed up the white-water system and stock temperatures rose, over a period of time, to 53°C. It was no longer possible to obtain the same sizing degree at the previous rosin/alum addition levels. The mill solved the problem by reducing the machine pH, to increase the charge on the alum species and by adding the rosin and alum close together late in the system.

Rosin sizing can be improved under hot stock conditions by the following methods:

- If the stock is hot inadvertently, add fresh water to cool it.
- If steam is injected, keep the point of injection prior to the size and as far away as possible.
- Run at a lower pH to maintain the charge on the alum.
- Use a polyaluminium species which maintains its cationic charge for a longer period at a higher temperature.
- Use a dispersed rosin size which is less sensitive than soap size to the chemistry of alum.
- Use a cationic rosin size which requires less alum.
- Improve the single-pass retention of rosin/alum by use of an efficient retention system.
- Add rosin/alum late in the system.
- Use a premix system or co-addition of rosin and alum to pre-form the rosin size precipitate before it attains the stock temperature.
- Add trim alum late in the system to fix unretained size precipitate.

Drying temperature

The effect of drying on sizing, using rosin size, is an important factor in obtaining the correct sizing. This can only be done utilising the correct drying procedure. The aluminium rosinate must reach the correct temperature, so that it is able to 'sinter', or melt and spread, to obtain the correct orientation on the fibres, giving the sizing degree required.

As has been previously explained, this differs for soap and dispersed sizes (anionic, or cationic), as soap size precipitates are almost entirely aluminium rosinate, which must have the correct moisture and temperature before sintering takes place, whereas dispersed sizes contain high levels of free rosin, which can melt redistribute and react with aluminium species under normal drying conditions.

High moisture and temperature in the early part of the dryers is not desirable, as the water removal in this section as steam, can disrupt size precipitates and cause size agglomeration rather than spreading. The optimum drying regime would be to slowly increase the temperature of the dryers until the middle, or just after the middle section, and then cool towards the end, as illustrated in Fig. 5.5. An example was seen of a mill which complained of lack of sizing, despite adequate size additions and with good retention figures. Paper from the press section was taken and dried in the laboratory under various conditions. It was found that using a high temperature on the pressed sheet gave a poor sizing effect. (The sheet was pressed on to a hotplate at 120°C.) Air-drying the sheet gave a poor sizing effect as did 2 min drying on a rotary dryer at 120°C (similar to the hotplate). However, drying for 4 min on a rotary dryer at 85–90°C gave excellent sizing.

As a result, the mill changed the drying profile to give cooler earlier dryers and slow gradient to hot dryers which then achieved more efficient sizing.

It should also be noted that pressing can have an effect on the drying profile required. Press moistures above 60% can also be detrimental to achieving full sizing development as can be seen in Fig. 5.6. There are no instant solutions to the problems of drying with rosin size other than those stated previously, as one mill system is different from another. It is very often a case of trial and error to get the best drying regime, as long as the basic rules are adhered to.

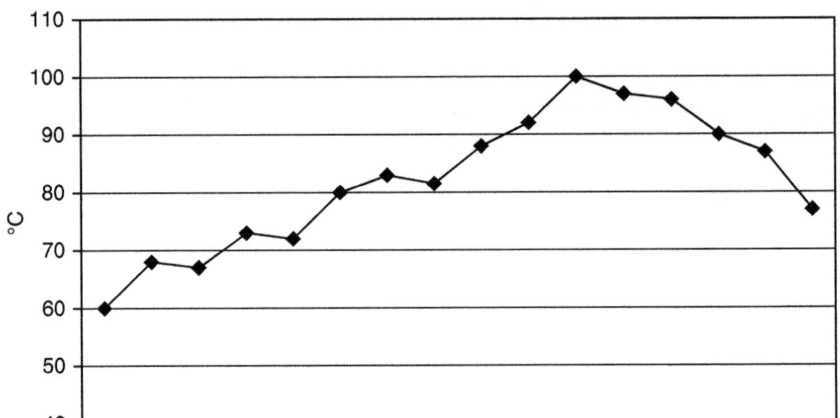

Fig. 5.5 A typical pre-size press paper temperature profile for dispersed rosin size

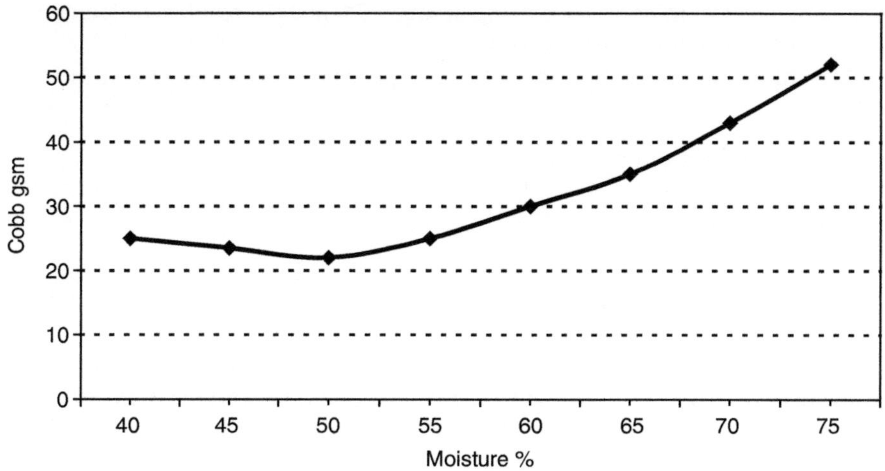

Fig. 5.6 Effect of press moisture on dispersed rosin size

5.4.1.3 Synthetic Sizing

The two major types of synthetic size employed at the wet-end of a paper or board machine, AKD and ASA, will be dealt with jointly as in many respects the wet-end chemistry is the same. Individual problems related to the sizes themselves will be dealt with later in this section.

Many articles and papers have been written regarding the reaction (or non-reaction) of AKD and the advantages of ASA over AKD and vice versa. The advantage of synthetic sizing over rosin sizing is also well covered in these papers and articles, and no attempt will be made to justify the use of one type of size or the other, and only solutions will be offered to possible problems that may occur.

Lack of sizing with synthetic size has, in many ways, the same causes as that for rosin sizes; it is just the solutions that may be different. There are obvious differences such as pH, but other factors such as those listed below are similar:

- Alkalinity
- Anionic trash (soluble anionic trash)
- System charge
- Stock temperature
- Drying temperature

pH

The synthetic sizes operate at a different pH range to rosin sizes (cationic rosin size systems can work at higher pH), in general between pH 6.0 and 8.5.

The natural pH of most systems being between 7.0 and 8.0 requires different operating conditions to that of rosin sizes.

The following effects may be seen on a paper or board machine at these pH values:

- Refining efficiency changes
- Drainage changes
- Retention changes
- Drying changes

Obviously, all of these have effects on the final sheet properties and can result in a different type of paper and board.

ASA and AKD react with cellulose under neutral or alkaline conditions, ASA being more reactive than AKD under these conditions and working more efficiently at a lower pH range than AKD. The hydrolysis rate is also quicker at high pH. So with ASA, a balance must be struck between hydrolysis and reaction. If the hydrolysis happens quicker than reaction, the sizing effect will be less. Also, if first-pass retention of size is not good, then at higher pH, e.g. pH 8.5, hydrolysis of the unretained size will occur. The hydrolysate of ASA is surface active in nature and acts contrary to the required sizing effect, but as previously stated, reacts with Ca^{2+}, Mg^{2+} and Na^{+} to give sticky compounds that deposit through the wet-end of the paper machine and in felts and driers.

With ASA the solution to the problem of hydrolysis is:

- Run at a pH between 6.0 and 8.0, preferably close to 7.0.
- Maximise first-pass retention.
- Add an aluminium compound to fix the hydrolysate (the aluminium salt of ASA is not a desizing agent and is less/non-sticky).

At lower pH (below 6.0) the reaction will be slower with cellulose and full sizing may not be obtained.

For AKD sizing, AKD is less reactive than ASA and thus is less subject to hydrolysis. It can therefore tolerate a much higher pH before hydrolysis occurs. It is also less reactive at pH 6.0–7.0 and therefore will not be as efficient at this pH range. The ideal pH for AKD is between 7.0 and 8.5, which is fortunate, as it is commonly used with calcium carbonate which buffers in the 7.5–8.0 region (depending on the calcium carbonate source).

However, if the pH does rise to 9.0–9.5, the rate of reaction will increase, as will the rate of hydrolysis. Thus, the same problems will occur as with ASA, except that the hydrolysate is not surface-active and desizing, so does not have the same adverse effect.

For AKD the solution is:

- Run at pH 7.0–8.5, preferably closer to 8.0.
- Maximise first-pass retention.

Below pH 6.0–6.5 reaction will be slow and it will be impossible to obtain the full sizing effect.

An example of the effect of pH on AKD sizing was seen at an integrated mill producing Kraft liner board. The mill was running at a pH of 7.5–8.0 with good sizing performance. The pH was controlled by addition of acid to the incoming stock. However, the stock pH increased to 9.0–9.5 when acid adjustment was not maintained. This led to an improvement in on-machine sizing, but with time the sizing efficiency started to decrease.
The sizing values obtained were:

pH 8.0 Cobb off-machine - 55 gsm
Cobb oven-cured - 23 gsm
Cobb natural-cured (24 h) - 22 gsm
pH 9.5 Cobb off-machine - 32 gsm
Cobb oven-cured - 32 gsm
Cobb natural-cured (24 h) - 33 gsm

Retention did not alter, and the size and retention aid flows were constant. The pH was decreased to 8.5, by acid adjustment, and the sizing values were:

Cobb off-machine - 43 gsm
Cobb oven-cured - 21 gsm
Cobb natural-cured (24 h) - 22 gsm

An example from ASA sizing is of a mill that was using ASA to produce wood-free papers using clay filler at pH 7.2. It changed to a calcium carbonate filler of a slightly lower surface area to try to improve sheet brightness, but at pH 8.3 instead of pH 7.2. It also expected to improve sizing efficiency. At the same level of filler addition/retention the sizing obtained was worse.

Analysis of the white water showed that approximately 15% more hydrolysed ASA was present, resulting in the reduced performance. The pH was lowered to 7.5 and sizing returned to previous levels. The improvement in sheet brightness was maintained and the expected improvement in sizing was seen.
For problems with synthetic size:

- Check the pH - has it changed?
- Is it too low?
- Is it too high?
- Adjust it to the optimum level.

Alkalinity

Alkalinity is normally thought to be a positive benefit for AKD and ASA sizing, as, for reasons previously stated, it can actively promote the reaction of AKD and ASA.

It does, however, depend on the source of the alkalinity. High levels of alkalinity from the hydroxyl OH^- ion will increase the pH to levels which are more favourable for hydrolysis. Such sources of alkalinity are NaOH and $Ca(OH)_2$. At high addition levels Na_2CO_3 can also affect the pH, increasing it to 9.0 and above.

The answer to the effect of alkalinity is to use small amounts of low-alkalinity additives. Sodium bicarbonate or calcium carbonate contain low levels of alkalinity (precipitated calcium carbonate can sometimes contain high levels of residual calcium hydroxide). Sodium carbonate contributes to higher levels.

Guidelines for alkalinity are:

- Check the alkalinity.
- What is the source?
- What is the level?
- Can it be replaced by a low alkalinity source if too high?

Then:

- Aim for 50–250 ppm $CaCO_3$ (as alkalinity).
- Try not to exceed 500 ppm $CaCO_3$.
- Use $CaCO_3$ if possible.
- Use $NaHCO_3$ rather than Na_2CO_3.
- Reduce by addition of fresh water if too high.

An example of problems with alkalinity is that of a mill producing lightweight coated paper with AKD. It was aware of the improvement in size cure by running at high pH and alkalinity and used Na_2CO_3 at the pulper, to give a total alkalinity of 500–600 ppm $CaCO_3$. Unfortunately, when paper was broked for any reason, and a large quantity of broke was used, the alkalinity increased to over 1,000 ppm and sometimes as high as 2,000 ppm. This resulted in loss of sizing efficiency, due to hydrolysis of AKD unretained on first pass.

This also led to loss of 'on-machine' sizing efficiency, and resulted in increased size levels being added. Problems were also seen with hydrolysed AKD deposits in the machine system.

The problems were solved by close control of the alkalinity and reducing the Na_2CO_3 additions when broke was added. The Na_2CO_3 addition was reduced by 1 kg/t for each 5% extra broke that was added which controlled the alkalinity at the correct level.

System charge

The most important factor for most paper and board machines with regard to sizing is the wet-end balance and the charge of the system. Synthetic sizes (and new rosin sizes) are mostly positively charged, or cationic, to enable them to be self-retaining in the wet-end. Unfortunately, there are many other chemicals at the wet-end that are also cationic and the fibre, fines and filler have only a finite cationic demand. If this demand is exceeded, then the cationic chemical being added will not be retained. Thus, you have to be aware of the various chemicals that are added and the charges on them, in order to not adversely affect the sizing efficiency. If the size is preceded by highly cationic material, which satisfies the cationic demand of the system, it will not be retained and will then build up in the white water.

This will, in turn, reduce the cationic demand of the stock still further and the lack of sizing will increase.
Examples of the type of polymeric cationic material which can give this effect are:

- Polyamines - drainage and retention
- Polyamideamine - wet strength
- Dicyandiamide resins - dye fixatives/promoters
- Cationic starches - dry strength/retention
- Polyethyleneimines - drainage and retention

It must be stressed that this applies to high levels of these types of polymers, and combinations of them. In normal use, the individual components should not give a problem. It is when they are used together at high levels, e.g. polyamine at, say, 2 kg/t, in combination with Polyamideamine at 20 kg/t in a system with a low cationic demand, i.e. low filler content, low fines content, pure virgin fibre.

Of course, you have to know the cationic demand of various components of a paper or board furnish. In general, the higher the surface area, the higher the demand, but this is not always a hard and fast rule. It can be said that:

- Virgin fibre has a higher demand than waste fibre.
- Coated broke has a higher demand than uncoated broke.
- Fillers have a higher demand than fines.

However, as most people are aware, waste furnishes contain a higher fines content than virgin fibre furnishes so the cationic demand of a waste furnish will be higher, although much of the waste material may have already been treated with cationic polymer or starch.

If the build-up situation occurs, it can be easily characterised in a paper or board machine by a series of symptoms:

- Persistent foam in white water
- Press picking
- Gradual reduction of sizing

The foam in the white water is caused by cationic material reducing the surface tension of the white water, enabling stabilised foams to occur.

Press picking is also caused by increased viscosity of the water-soluble macromolecules held in the sheet at the press which, under the hydrodynamic pressure, pulls off fines and filler. The increased viscosity is caused by excess cationic material.

The reduction in sizing is caused by loss in retention of cationic sizing agent, be it ASA or AKD.

It is not always certain which of these symptoms happens first, but press picking tends to be a forerunner of foam, as the system is becoming neutral to cationic, with sizing reduction occurring as the last effect. It should also be stressed that the sizing agents themselves can also contribute to the cationic nature of the system, particularly those containing polymeric cationic materials. So selection of the cationic ASA or AKD type is of equal importance to other polymeric materials being added.

An example of the problems of excess cationic material is a mill producing wallpaper backing from a ground wood furnish. The cationic material being added to the system (in order of addition) was:

10 kg/t dry cationic starch
20 kg/t wet strength resin (as received)
3 kg/t cationic promoter resin (as received)
30 kg/t cationic starch stabilised AKD size (as received)

Press picking quickly became a problem on changing to the grade, with a gradual foam build-up on the white water. Charge measurement was carried out on the system and was immediately found to be cationic. The promoter resin was turned off and fresh water added to the system. At the same time it was noted that the sizing degree was deteriorating, indicating lack of size. However, the removal of the promoter resin, and use of fresh water, restored the system charge to anionic, creating a cationic demand. The foam subsided, press picking decreased and sizing values were restored.

Unfortunately, the same mill decided to carry out ASA sizing, on the same grade of paper, at a later date. The ASA size used was a highly cationic polymeric cationic stabilised type. No promoter resin was used with the ASA, but the polymer in the ASA system was of medium charge density. The same result as previously was obtained – foam, press picking and loss of sizing. The mill now knew how to deal with this problem, but having no promoter to turn off they reduced the wet strength resin and the cationic starch, which affected the final paper properties and the retention. At a later date it changed to a starch stabilised ASA and the problems were solved.

Guidelines for system charge are:

- Maintain a cationic demand.
- Balance the cationic input.
- Check the system charge regularly.
- Know the charge of the chemical additives.
- Recognise the symptoms when problems occur.

There are several ways of charge measurement; all have their advantages and disadvantages. The main ways of measurement are:

- Particle charge detector (PCD or Mutek™)
- Zetameter™
- SPM™ (streaming potential meter)
- Cationic/anionic demand titration

Obviously, the cheapest is the titration method, but the best is that which is continuously measured on-machine. All of the instrumental procedures are capable of this – but at a cost.

However, having any means to check charge balance and knowing what good and bad values are, is the best means of knowing when things are going wrong.

Anionic trash (and soluble anionic trash)

Anionic trash, in synthetic sizing, has a similar effect on size retention as that for rosin size, except that in this case, it is the size that carries the cationic charge, not the alum. This will mean that it is the sizing agent that will be adsorbed onto the anionic trash. If the size is adsorbed onto the anionic trash, sizing will be subject to the retention of this anionic trash and will vary accordingly. This is why much more emphasis is placed on the correct retention system to improve anionic trash retention, or fines/filler retention, as much of the size can also be adsorbed on this.

An example of the amount retained on anionic trash, from a 100% waste furnish, has shown that 85% of a cationic AKD was retained on the fines fraction. Thus, in this case, fines retention was extremely important for lack of sizing. In order to counteract this effect, it is preferable to use an ATC prior to size addition. These ATCs are cationic polymers or Al^{3+} species. Typically, they should be added to the source of anionic trash, i.e. if the source is from the pulp, they should be added to this. If the source is coated broke, the ATC should be added to the broke. If the source is white water, the ATC should be added to this.

However, many mills use dual component retention systems such as Compozil™, or Hydrocol™, which involves the use of a cationic polymer or starch prior to addition of anionic component. This cationic component can sometimes be used as the ATC part, and size added after this addition.

Some sizes also contain additional cationic polymer, which, as it is more mobile and effective than that of the size/polymer particle, acts as an in situ ATC. However, these systems are slightly less effective than separate ATC addition. If ATCs are not used, a lack of sizing will always be the result, or certainly a loss of sizing efficiency. With soluble anionic trash in integrated mills, or present in pulp, the same holds true as for anionic trash, except that in this case the sizing agent will be totally lost to the system, and will not be retained on the anionic trash or fines. This will obviously affect the sizing efficiency dramatically. The type of ATC to use is more important in synthetic sizing, as the system charge balance is important and excessive cationic material can be just as detrimental as the anionic trash.

Levels of cationic polymer must be carefully determined in order to avoid this happening. Typical ATCs are:

- Polyamines
- PolyDADMAC
- Polyaluminium chloride
- Aluminium chlorohydrate
- Alum

But other cationic polymers used for other purposes will often act as ATC, as previously mentioned, such as:

- Cationic starch (strength/retention)
- Cationic polyacrylamides (retention)

- Cationic polyamideamine (wet strength resins)
- Cationic dicyandiamide resins (dye fixatives/promoters)
- Cationic polyethyleneimines (drainage)

Of these types, cationic polymers are more effective at lower levels of addition than the aluminium compounds (they are more expensive), but aluminium compounds have the benefit of losing their cationicity quickly, especially alum at pH 7.0+, so will not affect the system charge as much at high addition levels. There is a possibility with alum, of deposits of basic alum ($Al(OH)_3$ - aluminium hydroxide), but if there is careful control this can be minimised.

Typical levels of addition would be 50–200 g/t (active) for the cationic polymers and 500–1500 g/t for the aluminium compounds.

However, a guideline for the use of ATC type would be to treat 'like with like' i.e. if the anionic trash is low molecular weight soluble highly anionic material it should be treated with a low molecular weight soluble highly cationic material, e.g. alum/PAC and high molecular weight low charge anionic trash (fillers/fines) treated with similar cationic material (polyacrylamides/starch).

An example of the effect of anionic trash giving lack of sizing is of a mill using cationic ASA to size waste-based linerboard. The mill was using a cationic polyacrylamide for retention, but sizing varied considerably, requiring frequent size level adjustment. They changed to a dual component system with cationic starch (also added for dry strength properties), the ASA addition point was moved forward of the cationic starch and sizing became much less variable and more predictable.

The solutions for anionic trash with synthetic sizes are similar to those of rosin.

- Establish the source.
- Treat with ATC or polymer.
- Move size addition to after ATC.
- Check system cationic demand/system charge.

Stock temperature

Stock temperature is, surprisingly, no more important for ASA sizing than for AKD and neither is as affected as rosin sizing (no alum chemistry/temperature problems).

The AKD and ASA particles are cationic in nature and covered (see Fig. 5.7) by a layer of colloid (starch or polymer) which protects them from the harsh realities of hot stock conditions. As long as the protective layer is not broken and the particles are still firmly fixed to the fibres/fines, they will remain unaffected. However, at temperatures above 60–70°C the particles begin to lose stability. For AKD, this is above its melting range of 45–60°C, and similarly for ASA, where the ASA begins to become more mobile within its particle. If the particles break, then the AKD and ASA are exposed to heat and water at alkaline pH, which hydrolyses them extremely rapidly – ASA more rapidly than AKD.

However, to avoid this, some of the same conditions hold for ASA and AKD, as for rosin sizes.

Fig. 5.7 The size particle

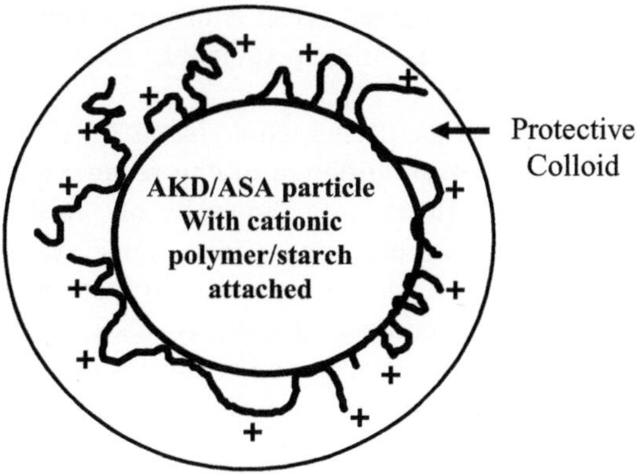

- If stock is hot inadvertently, cool by addition of fresh water.
- If steam is injected, keep the injection point as far away from the size.
- Add size late in the system.
- Improve the single-pass retention of size by use of an efficient retention system.

Drying temperature

For synthetic sizes the action of the drying cylinders is different to that of rosin/alum, where the main function is to spread and fix the rosin. With AKD and ASA sizes, the spreading of these is not as important, as AKD is a low melting point wax, and ASA is an oily liquid. The main purpose of the drying section is to break the AKD or ASA particle, as shown in Fig. 5.7, to allow the AKD or ASA to spread and then react.

However, the drying requirements of AKD and ASA sizes are vastly different, for the following reasons:

- AKD is slow to hydrolyse and react.
- ASA is fast to hydrolyse and react.

If ASA is subject to heat and moisture at the same time, it will hydrolyse and react at similar rates, resulting in reduced efficiency. Therefore, it should be dried similarly to rosin size, but with sufficient heat to break the ASA particle at a low moisture content.

For AKD, as it is much slower to hydrolyse and react, it should be dried quickly, reaching a high paper temperature early in the drying section, to enable the AKD particle to break. The AKD is sufficiently slow in its reaction with moisture in the paper, that it is not hydrolysed. It will then be able to react with the cellulose fibre, after spreading, in the later cylinders, as long as the temperature is above the melting point of the AKD itself. The reaction of AKD, however, as it is slow, really requires a greater energy input to occur than that of ASA.

Fortunately, unreacted AKD is also a sizing agent, albeit less efficient than reacted AKD and is often responsible for the on-machine sizing effect seen from AKD sizing.

An example of the drying requirements of ASA and AKD is shown in Fig. 5.8.

An example of the different drying requirements of AKD giving problems of lack of sizing efficiency is shown by one mill. The mill produced paper for off-machine coating and used a joint sizing system at the wet-end of ASA and AKD. The use of ASA was to enable off machine sizing to be obtained, to enable immediate coating, compensating for any lack of off-machine sizing from the AKD. However, the drying profile was set up for AKD sizing and the mill complained about the low efficiency of the ASA, at an addition level that should have obtained the sizing degree required alone.

It was noted, that after several months, deposits of hydrolysed ASA were found in the pre-dryers and on the dryer frames. This led, unfortunately, to a fire in the dryers. It was later found to be caused by excessive heat and moisture in the early dryers (required for AKD) hydrolysing the ASA and steam distilling it onto the dryer frames. The dryer profile was altered to be more favourable to ASA, the problem was removed and sizing efficiency of the ASA improved.

As stated for rosin size, the drying temperature is important for synthetic sizes, but the profile is more important and as long as a profile similar to that of the ideal is used, lack of sizing should not be a problem. However, it is a case of trial and error to obtain the correct dryer profile for individual machines, to obtain the best balance of sizing performance and energy input.

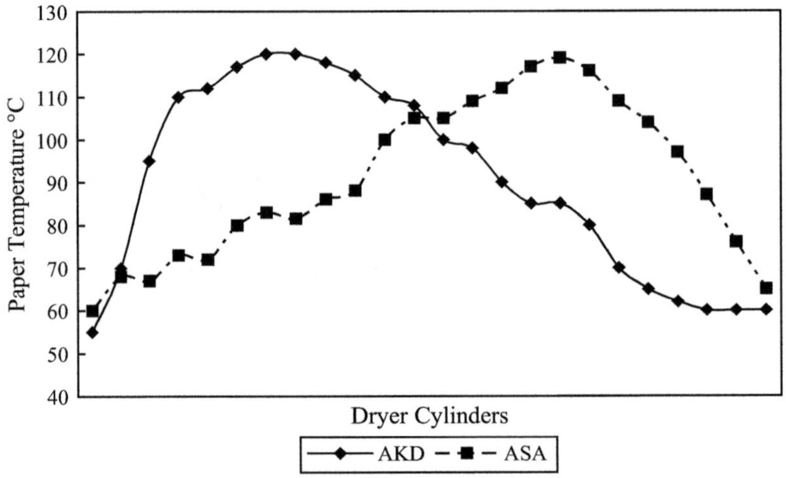

Fig. 5.8 Pre-dryer paper temperature

5.4.2 Slow Cure

5.4.2.1 Rosin Size

Slow cure of rosin size is not a factor that should occur. If rosin size and alum are present in a furnish in the correct ratios they will give the sizing effect immediately. If the ratio is incorrect, or if the drying is incorrect, then there will be no sizing effect.

5.4.2.2 Synthetic Size

Slow cure with synthetic sizes is the same for both ASA and AKD (and all other synthetics). It can normally be related to the drying profile and the moisture of the paper or board. The paper or board has to reach a certain moisture content and temperature before the size particles will break and align/react. If they do not reach the values they will not give the true sizing effect.

However, if they are cured in the oven, the size will be able to spread and react. This is the so-called oven-cured size value, which shows that the size is retained by the sheet. This cure effect is enhanced by calendering, rewinding and high temperatures in the reel. All of these exert temperature on the size particles, enabling them to react in the paper or board after production.

Slow cure is normally seen in high moisture papers and boards and the effect will be greater for AKD than for ASA, because of the slower reactivity of AKD and the difference in melting points, i.e. a liquid versus a low melting point wax. It is possible to improve the curing of AKD (and ASA) by running to a lower moisture <4% prior to a size press/film sizer and subsequently drying less after the size press/film sizer, but this may impact on the surface properties.

An example of a mill experiencing slow cure was a board machine converting to alkaline size. It ran for some time with AKD size, before converting to ASA, as the board did not cure quickly enough.

However, with ASA, it also experienced a 'cure down' effect that was unexpected. To remedy this it improved the drying profile and was able to run at a lower residual moisture content on machine, prior to using water calenders, to increase the moisture to customer specification.

To avoid slow cure:

- Run at low moisture (less than 7%).
- Improve drying profile.
- Reel up hot.
- Improve drainage at wet-end.
- Decrease ex-press moistures.

5.4.3 *Loss of Sizing, or Fugitivity*

Loss of sizing, or fugitivity, is only seen when a paper or board has an 'off-machine' sizing value which deteriorates with time. It has often been confused, especially with AKD sizing, with failure to develop sizing, or not naturally curing to the same values obtained by oven curing.

5.4.3.1 Rosin Sizing

With rosin size, fugitivity can be caused by two contributory factors:

- Moisture
- Lack of alum or aluminium species

Both of these factors are needed to give fugitive sizing and it is only ever seen in high moisture paper and board. The reason for it occurring is, when due to a lack of Al^{3+} attached to the fibre, the rosin is only very loosely attached to the fibre and fines. Moisture can then disrupt the alignment of the rosin reducing the hydrophobic effect. For soap sizes, this is more likely to occur, because of higher alum requirement and poor formation of aluminium rosinate. It occurs most often with high-soluble anionic trash systems, as the alum, or aluminium species, is lost to the system in solution. This is more likely at higher pH, as the alum loses its cationic charge and thus, is less likely to occur with polyaluminium species which maintain their charge at this pH.

An example of this fugitivity was seen in an integrated board mill, using ground wood to produce coated board. The mill was using dispersed anionic rosin size and had reduced the alum to a minimum. However, during the on-machine coating operation, a high moisture situation was produced, which was effectively sealed into the board. At times, when the soluble anionics were high, the combination produced sized board which gradually became unsized with time (when it reached the customer). The mill solved the problem by improving the drying prior to the coating station and running a higher level of alum.

The situation of fugitivity with rosin size in paper is rarely seen and this was instantly diagnosed as due to the two factors.

5.4.3.2 Synthetic Sizing

With the synthetic sizes, the problem of fugitivity came to the fore in the USA with PCC. The effect was seen initially in the USA/South America because of the quality and availability of GCC and chalk and the process for PCC was introduced in order to allow the conversion to alkaline of fine paper machines.

During the precipitation process, it is possible to entrain $Ca(OH)_2$ in the $CaCO_3$ filler particle. This $Ca(OH)_2$ is the source of the fugitivity, as it can leach from the particle over a period of time and hydrolyse the size particle attached.

ASA and AKD sizing agents form a sized layer on the PCC filler surface. With ASA this happens rapidly and there is no residual ASA available to react as any residual is hydrolysed. However, with AKD this reaction is slower and there is 'unreacted AKD' on the PCC surface. This 'unreacted AKD' gives a sizing effect on the filler particle (~5 times less efficient than reacted AKD).

As the $Ca(OH)_2$ leaches from the particle and comes in contact with the 'unreacted AKD', it creates, with the moisture in the paper, a high pH environment, which can cause hydrolysis of the unreacted AKD and loss of the sizing effect on the PCC particle. This is then able to adsorb water, or ink, and create a desizing effect, although the fibre itself may still be sized. Obviously, the higher the entrapped lime level and the higher the filler content, the greater the effect.

Evidence for the leaching of lime in PCC slurry has been seen:

- pH PCC slurry as made - 8.0
- pH PCC slurry 1 month - 10.6
- pH extract of paper 1 month (fugitive sizing) - 10.3
- pH extract of paper 1 month (non-fugitive made with ground CaCO3) - 7.8

The $Ca(OH)_2$ content of the water extracted from a PCC slurry has also been determined:

- $Ca(OH)_2$ as received - 0.17 g/l
- $Ca(OH)_2$ 1 month - 0.48 g/l

It appears that this leaching effect (and thus hydrolysis of AKD size on PCC) is accentuated by highly cationic material.

The rate of hydrolysis of AKD emulsions of varying cationic charges which were mixed and held at ambient temperature in a PCC slurry of known $Ca(OH)_2$ content has been measured and appeared to support this theory:

- Very low cationic AKD - 0.78 g/day
- Low cationic AKD - 0.91 g/day
- Medium cationic AKD - 2.11 g/day
- High cationic AKD - 2.87 g/day

In comparison, a sample of the same PCC was then treated with excess carbon dioxide (CO_2) to further remove the $Ca(OH)_2$ entrained and the result was:

- Medium cationic AKD - 1.42 g/day

This is caused by the higher cationic charged AKD having a greater affinity for the hydroxyl (OH^-) of the residual lime. The problem could have occur to the same extent with ASA but, due to its extreme reactivity, it would have already reacted, or hydrolysed, before the $Ca(OH)_2$ could leach from the PCC.

The problem with AKD is that this process occurs over a period of up to 1 month; depending on the moisture content of the paper and ambient temperature (higher moisture/temperature is bad). The effect can be camouflaged by use of sacrificial levels of AKD, but this will result in higher size levels. There is also a lesser form of this fugitivity which is called sizing regression. In this case, the sizing level is obtained but reduces to a lower, but still sized, level.

This effect has only been seen in PCC produced in the slurry form. Solid PCC which has been obtained by drying the slurry does not show this fugitive effect, to the authors' knowledge. Until now, the producers of PCC have optimised their processes to minimise the entrapment of Ca(OH)2, and AKD suppliers have modified their formulations, such that it is possible to use AKD in PCC-filled paper. However, the improvements in ASA application mean that ASA can be a more economic solution. The solutions for fugitive sizing with PCC slurry appear to be:

- Separate PCC addition from highly cationic addition.
- Keep the cationicity of the system low.
- Use 'sacrificial' levels of size.

But more recently the problem has also been solved by:

- Use of surface size to mask the effect

High moisture

Fugitivity with high moisture is a problem specific to AKD sizes. It only occurs at very high moisture levels ~8–10% and tends to be specific to board manufacture. In this case the 'unreacted' AKD is again the culprit. As most of the initial sizing effect of AKD is due to the non-reacted AKD under poor drying conditions, the 'unreacted' AKD needs to react (cure) to give a permanent sizing effect. However, it can be hydrolysed under high moisture conditions at above pH 7.5. The AKD then becomes non-sizing and a fugitive effect occurs.

An example of this was seen in a board machine which pressure laminated the board directly from the machine. The board was produced at 5–7% moisture and the pressure lamination added a further 3–4%. Board produced was analysed and it was found that a low level of reacted AKD was present for the sizing effect obtained, but a high level of total AKD was present. Board produced at 6–8% total moisture was found to give stable sizing results with time, whereas board at higher moisture content gave a fugitive sizing effect:

Total AKD in board 6–8% moisture content - 0.34%
Total AKD in board 11–12% moisture content - 0.35%
Reacted AKD in board 8–9% moisture content - 0.12%
Reacted AKD in board 11–12% moisture content - 0.05%

Initial sizing, 6–8% moisture	5% pickup, 24 h immersion
Initial sizing, 11–12% moisture	4% pickup, 24 h immersion
1 month sizing, 6–8% moisture	4% pickup, 24 h immersion
1 month sizing, 11–12% moisture	20% pickup, 24 h immersion

The solutions to this are:

- Reduce moisture content of board to below 7%.
- Improve reactivity of AKD (drying, pH).

5.4.4 Deposits

Deposits are a natural result of problems at the wet-end of a paper or board machine. They are normally related to problems of retention of the sizing agent although they can be caused by incompatibility. They may occur in the stock approach system, on the wire, on the presses or in the drying section. Many of the reasons for deposits have already been covered, but this section will try to summarise them.

5.4.4.1 Rosin Size

The main reason for deposits with rosin size is due to the presence of alkalinity or high pH. The Na^{+} Ca^{2+} and Mg^{2+} caused by the alkalinity react with rosin size, forming small flocs which are no longer cationic in nature, as stated previously. They are thus less likely to be retained, but grow to form macroflocs which can be deposited in the stock approach system on the wire. These macroflocs being organic in nature are often associated with oil from defoamers and antifoams and, of course, with any fillers that are in the system.

Deposits can also occur at high pH due to a lack of Al^{3+} species with alum. Alum may have reacted with rosin, but there is not sufficient charge on the particle to retain it. It then will be deposited in the stock approach system as basic aluminium rosinate. This is aluminium rosinate where the cationic part of the aluminium has been neutralised by the hydroxyl ion. This tends to occur more with soap sizes than with rosin sizes as the alum needs to react with the rosin soap and then be retained. With rosin dispersions this effect is less.

A further cause of deposits is a lack of alum. If sufficient is not added to the system to precipitate the rosin, it will deposit in the whitewater and thin stock system.

All of these effects are enhanced by other effects, such as high stock temperature. The lack of alum may also cause deposits at the presses and dryers, as the size precipitate will not be firmly anchored to the fibre and fines and can be removed, at both press and dryers, with fines and fillers. If there is also a chemical imbalance in the system, the fines and fillers can be removed at the presses. In this case, there will always be a high level of fines/filler to rosin ratio.

Finally, deposits occurring in the dryers can be due to poor drying profile, with excess heat in the early dryers causing size agglomeration. This will coat the dryer surface and felts and cause sticking or, by the steam removal of the precipitates, give the same effect. However, it is this same effect that makes the use of rosin for MG, or Yankee, machines as the rosin causes a sticky surface on the MG cylinder, enhancing the sheet adhesion and subsequent gloss.

5.4.4.2 Synthetic Sizes

The reasons for deposits with synthetic sizes are much the same as for rosin, although the scale of the deposits tends to be smaller due to the lower levels of chemicals employed and, in general, better retention of these sizes.

The main reason for deposits with synthetic sizes is poor retention which may be caused by system imbalance and subsequent hydrolysis, or breaking, of the cationic emulsions. If the system is in balance, up to 90% of the size will be retained on the first pass through the system and the majority of the remainder on the second pass.

However, if there is a wet-end imbalance, less will be retained on the first and second passes and it will be subject to hydrolysis in the stock system. This is more of a problem with ASA than AKD, as the hydrolysis product is also very sticky, which can lead to deposits in stock approach systems and in wires and felts. AKD, being less prone to hydrolysis, may remain in the system for a much longer period, before being hydrolysed, but will still give deposits in the wet-end, although they are less harmful than ASA deposits.

Conditions which may destabilise the particles should be avoided, such as high temperature and high shear. If the size particles are stabilised before they are retained then, for AKD, pure AKD may be deposited in the stock approach system, whilst ASA will be hydrolysed rapidly and may cause sticky deposits in the stock approach.

Poor retention of fillers and fines caused by the poor retention and/or wet-end imbalance can also result in deposits of filler/fines/size at the presses and press felts and pre-dryer doctor blades.

Dryer deposits with ASA are caused by hydrolysis and steam distillation of the oily ASA by excessive early dryer temperatures as mentioned previously.

There have been cases of incompatibility, e.g. OBA and size added at the same point, causing destabilisation of size (and OBA), but this would be normally easy to spot and rectify.

To solve deposit problems:

- Identify the type of deposit (ASA salt, pure AKD, Na rosinate).
- Establish the cause (retention, destabilisation).
- Check the system (charge, addition, sequence).
- Take action as identified in previous sections.

5.4.5 Other Problems

5.4.5.1 Two Sidedness

This is a fines/filler retention related problem. Fines/filler are removed from the sheet on the wire and as they have size attached to them the size will be proportionally greater in the top side of the sheet. This will give harder sizing on the top side. The solution is to improve the filler/fines retention. This may involve an ionic imbalance, use of a more efficient retention system, or even reduction of wire vacuums and foils.

5.4.5.2 Size Migration

This can occur with synthetic sizes when the size particles have not been broken or in high moisture situations, only in multiply paper or board. It tends to occur where one ply is hard sized and the remainder are not. It occurs mainly with AKD, but can occur with ASA if the particle has not been broken. It can be reduced or eliminated by:

- Gradient sizing of plies (i.e. hard/med/soft)
- Maintaining the same ionic charges in plies
- Reducing the moisture content
- Better drying

5.4.5.3 Slip

Slip is not a sizing problem, as such, but is caused by the use of excess AKD and subsequent hydrolysis of this excess material. The hydrolysis product of AKD, the so called stearone, has a higher melting point and is a more slippery material. The hydrolysis of AKD also tends to occur on the surface of paper, rather than internally, which means there is always an excess of stearone at the surface. This means that there can be slippery surfaces in contact causing this problem. It should be stressed that this only occurs with excess AKD. In general levels of AKD below 1.5 kg/t in fine paper, 2.0 kg/t in recycled paper and 3 kg/t in waste-based board should not give this problem. It can be reduced, or eliminated by:

- Use of a more efficient retention system
- Optimising AKD retention
- Improving AKD reaction
- Reducing the moisture content
- Combination sizing (optimised AKD at the wet-end and surface size at the size/film press)

5.4.5.4 Foam

Foam, like slip, is not a sizing problem as such, but a symptom of other problems which may affect sizing. These may be a cationic imbalance with synthetic sizing, caused by excess cationic material, or calcium carbonate loaded waste reacting with alum to generate CO_2. It can be reduced by eliminating air leakage into stock at pumps and seals and also use of defoamers, but it is better to tackle the root cause.

5.4.5.5 Defoamers

Defoamers can adversely affect sizing agents by adsorbing them, especially if they are oil- or kerosene-based products. They also contain surface active agents, which if used at extremely high levels, can give a desizing effect to paper and board, although this is extremely rare.

5.4.5.6 Shelf Life

All products supplied as sizing agents have a finite shelf life and it is surprising how many times problems with size efficiency have been caused by using a product outside its shelf life. This should always be checked if no obvious cause can be seen on the machine.

5.4.5.7 Size Storage

As with the shelf life, it is remarkable how often material that is marked 'Protect from extremes of temperature' is left outside, exposed to the elements, where it can deteriorate in quality. Good housekeeping with respect to chemical storage can eliminate this as a problem.

Acknowledgements Thanks are given to Mr I. Hiskens and Mr R. Britton for their invaluable contributions to the rosin sizing sections of this troubleshooting guide.

Additional thanks are given to Mr. P. Otway for his section in the first edition of this book, which inspired part of this contribution.

Chapter 6
Fillers

Leslie McLain and Danny Ingle

6.1 Introduction

The American Heritage dictionary defines filler as 'something added to augment weight or size or fill space'. Historically, commercial papermakers have used a variety of inexpensive, minimally beneficiated minerals as fillers for economic extension of more costly wood fibre. As such, these fillers played a relatively inconsequential role in contributing specific quality characteristics to the final sheet. However, as paper grades have evolved, the role of mineral fillers has dramatically expanded to contribute specific functionality to final paper grades. In general, this has resulted in a broader offering of mineral products to the papermaker delivering a range of optical and physical properties. Additionally, the use of mineral fillers may significantly impact dynamics on the paper machine itself. For example, the type and level of filler can dramatically affect chemical demand, drainage, speed and drying rates. A basic understanding of the fundamental characteristics of fillers and their resulting impact, both within the paper matrix and on the paper machine, is a critical requirement for cost-effective grade optimization.

In simplistic terms, paper may be considered a blend of materials primarily comprising cellulose fibres, minerals and air. The interaction of these materials with one another significantly influences the physical properties of the resulting paper web. Similarly, the interactions of these materials with one another and with light are the primary factors impacting scattering and absorption coefficients which often control brightness and opacity development, a key reason for using pigment in paper.

Though there are some characteristics unique to specific pigment types, in general, mineral fillers can be universally described using a set of fundamental characteristics. By categorising fillers in this manner, paper producers may predict the behaviour of filler in the presence of fibre, air and light of varying wavelengths, and are thus

L. McLain
Technical Manager – Uncoated Applications, Imerys, Pigments for Paper,
North America, Roswell, GA, North America

D. Ingle
Technical Marketing Manager, Asia Pacific, Imerys Asia Pacific PTE Ltd.,
80 Robinson rd 19-02#, Singapore 068898

I. Thorn and C.O. Au (eds.), *Applications of Wet-End Paper Chemistry*,
DOI 10.1007/978-1-4020-6038-0_6, © Springer Science+Business Media B.V. 2009

better prepared to choose effective pigments for any given application. It should be noted, however, that fibre type, degree of refining and point of filler addition will all have an impact on the final sheet properties.

6.2 Types of Fillers

Globally, kaolin clays and calcium carbonates are the most commonly used fillers on a volume basis. Kaolins may be subdivided into two broad classes, hydrous or calcined, while calcium carbonate may be subdivided into chalk, ground or precipitated. The size and shape of these pigments can vary widely depending upon the targeted paper application. Less commonly used fillers include talc, titanium dioxide (TiO_2) and amorphous silicates. These smaller volume fillers are often considered premium pigments as they tend to contribute a higher degree of functionality to the sheet, and are often significantly more expensive than traditional kaolins and carbonates. Filler type and loading can vary widely based on geographical, economic and wet-end chemistry considerations along with final sheet performance requirements. A brief description of the major filler classes follows.

6.2.1 Kaolin Clays

Kaolin (or clay as it is commonly referred to in the paper industry) is a hydrated alumina silicate with the chemical formula $Al_2O_3{\cdot}2SiO_2{\cdot}2H_2O$ comprising 39% Al_2O_3, 46% SiO_2 and 13% H_20. The vast majority of commercially mined kaolin clays are found in England, Germany, the Czech Republic, southeastern USA and Brazil, though there are also deposits of economic interest in the Ukraine, China and Australia. Sites are classified as either primary or secondary based upon the manner in which the kaolin has been formed. The majority of sites mined in Europe are considered primary, which is to say, the clays were formed by in situ hydrothermal alteration of granite and feldspar, and are often intimately bound with granite and quartz. Secondary kaolin deposits in southeastern USA and Brazil were formed by the weathering and erosion of granite and feldspar. The kaolin mineral was then transported by river and deposited sometimes hundreds of miles from the originating source. Secondary deposits tend to yield a higher recovery and are often found in bands of somewhat uniform particle size, due to natural sedimentation.

Kaolin clays may be subdivided into two broad classes: hydrous and anhydrous (more commonly referred to as calcined).

6.2.1.1 Hydrous Kaolin

Hydrous kaolins (i.e. kaolins containing ~13% water of crystallisation based on mass) were first commonly used as papermaking fillers in the nineteenth century

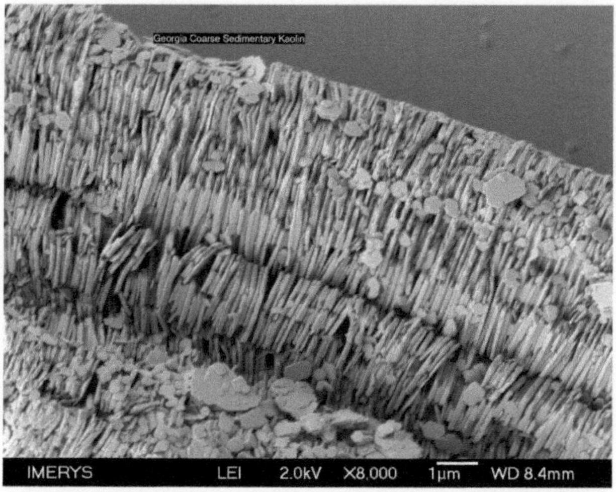

Fig. 6.1 Georgia, US crude kaolin

and were the largest volume fillers until the 1980s, when alkaline papermaking was widely implemented.

Hydrous kaolins are naturally occurring minerals found as hexagonal plates of varying crystalline uniformity. The plates are often associated with one another as 'stacks' or 'booklets' (Fig. 6.1). These particles have low aspect ratios and typically exhibit small surface area. Aspect ratio is defined by the ratio of the plate diameter to the plate thickness.

Hydrous kaolin products are differentiated based on particle size, brightness and aspect ratio, or degree of platiness. The wet processing routes for most hydrous kaolins include screening, particle size classification and magnetic separation. The objective is to remove large impurities such as quartz and to improve the brightness of the crude by removing iron containing impurities. Additional processing steps may include delamination, by which the stacks of plates are cleaved from one another, thereby increasing surface area and degree of platiness, or aspect ratio (Fig. 6.2). Additional brightness gains may be achieved through ozonation, flotation and oxidative and/or reductive bleaching. Kaolin clays may be shipped to paper mills in dry form with little or no dispersant associated with the particle, or in dispersed slurry form at a typical solid range of 60–70%.

As stated earlier, hydrous kaolins were traditionally used for fibre extension, but, depending on the degree of beneficiation, these fillers may offer a range of optical and physical benefits to the final sheet. At the time of this writing, it is estimated that 1.8 million tonnes of hydrous kaolin is consumed annually as filler for paper applications on a global basis. The largest consumers of hydrous kaolin fillers are supercalendered paper producers, which are located primarily in Western Europe, Scandinavia, Finland and North America. The required attributes of the final SC sheet include high gloss and good printability, characteristics which are predominantly attributed to the use of kaolin.

Fig. 6.2 Delaminated kaolin

Fig. 6.3 Calcined kaolin

6.2.1.2 Anhydrous or Calcined Kaolin

Calcined kaolins are sometimes referred to as thermally engineered kaolins, and are formed by passing hydrous feed clays through a calciner at temperatures ranging from 800° to greater than 1,000°C. The intrinsically bound water of crystallisation is driven off resulting in agglomerates of fused kaolin plates with a relatively high degree of internal void volume as shown in Fig. 6.3. This particle shape results in increased light refraction and scattering coefficient thereby yielding significantly higher sheet opacity and brightness compared to hydrous kaolin grades. Additionally, the combination of high surface area and void volume contribute to increased coefficient of friction and reduced rate and degree of ink penetration in lightweight wood-containing grades. Ink penetration is minimised by trapping ink within the mineral pores thereby immobilizing the ink near the surface of the sheet. Potential negative attributes of calcined kaolin include deleterious impact on strength and higher pigment abrasion associated with the calcining process. For these reasons, calcined kaolins are used in relatively small amounts where light scatter and print-through control is critical. On an annual basis, approximately 100,000 tonnes of calcined kaolin is used globally, primarily in newsprint and speciality wood-containing grades produced in North America (e.g. Directory, LWC and SC roto grades).

6.2.2 *Calcium Carbonates*

During most of the twentieth century, the majority of papermaking processes were carried out under acidic conditions. This largely limited the use of calcium carbonate (chemically expressed as $CaCO_3$), a mineral that is unstable at acidic pH. However, growing pressure to increase permanence in archival papers triggered a widespread trend among freesheet producers to alkaline papermaking in the 1980s. This chemistry change allowed papermakers to capitalize on the high brightness of calcium carbonate products which have since grown into the largest pigment class used as filler in woodfree production.

In wood-containing specialities, however, wet-end chemistry issues remain a significant hurdle, though carbonate growth is increasing in these markets. In most cases, pH must be controlled within a specific neutral range (typically 6.8–7.5) in order to avoid severe alkaline darkening. There are a number of emerging methods employed to control pH and avoid dissolution of calcium carbonate, including surface treatment of the carbonate particle itself. However, these neutral pH ranges offer an optimal environment for bacterial growth and pitch deposition, so serious consideration must be given to wet-end chemistry needs for specific applications to determine the optimal pH range for individual mill sites.

Calcium carbonate in its most common form is limestone and occurs naturally throughout the world as fossiliferous deposits composed of shells, teeth and bones from prehistoric organisms. Calcium carbonates used in the papermaking process fall into three groups: chalk, ground (GCC), or precipitated (PCC).

6.2.2.1 Chalk

Chalk is a variety of limestone mostly comprising fossilized shells from aquatic life forms. It tends to be softer than the limestone used to produce ground calcium carbonate. The most commonly utilised deposits are found in England and Denmark. Chalk is not found to a large extent in North America or Asia, so as a result the use of chalk in paper filling applications is mostly restricted to Europe. Chalk is most often produced using a dry processing route and is delivered to the mill site in dry form for makedown just prior to use.

6.2.2.2 Ground Calcium Carbonate (GCC)

GCC is derived from marble, and may be geologically described as a metamorphic limestone that has re-crystallised to form very hard deposits. These marble crudes can be physically ground into particles small enough for use in the papermaking process and are widely referred to as ground calcium carbonate. Processing steps for most GCC filler products include screening, grinding and particle size classification as well as some flotation for brightness enhancement. Once processed, these products are typically anionically dispersed in slurry form and distributed from large production facilities via truck and rail to mill sites. GCC products exhibit low aspect ratios and are blocky pigments of varied size (Fig. 6.4). Like hydrous kaolins, GCC products tend to allow relatively high fibre extension as the low surface area coarse particles are not as deleterious to fibre–fibre bonding. In 2006 approximately 2.1 million tonnes of GCC were used for filling applications globally, with a fairly even distribution of use throughout the major papermaking geographic regions.

6.2.2.3 Precipitated Calcium Carbonate (PCC)

Finally, limestone may also be converted to a range of synthetically precipitated calcium carbonates through specific reactions of milk of lime and carbon dioxide. In most cases, the limestone (or a treated derivative thereof) is shipped to a satellite plant located adjacent to an integrated paper mill where it is converted to PCC, exhibiting very uniform particle size and specific crystalline shapes. The process makes use of carbon dioxide from the pulp mill flue gas, thereby reducing overall carbon emissions from the papermaking facility. By altering reaction rates, temperatures and chemical additives, a wide range of particle sizes and shapes can

Fig. 6.4 Ground calcium carbonate

Fig. 6.5 Aragonite precipitated calcium carbonate

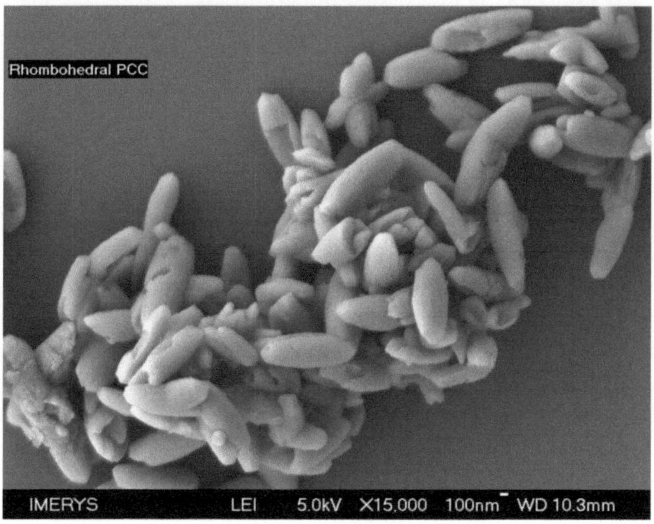

Fig. 6.6 Rhombohedral precipitated calcium carbonate

be achieved depending upon the specific performance requirements desired, and can therefore be customised for the host mill (Figs. 6.5, 6.6 and 6.7).

Because PCC is most often produced in close proximity to a host paper mill, filler products are usually delivered to a mill storage tank via pipeline at low solids. This eliminates the need for dispersing agents, which almost always interfere with papermaking retention chemistries.

Fig. 6.7 Scalenohedral precipitated calcium carbonate

With the advent of alkaline papermaking, PCC use exploded in the 1980s and 90s and, as a result, PCC is the largest mineral used in woodfree production on a global basis, with approximately 2.4 million tonnes consumed annually.

6.2.3 Titanium Dioxide

Titanium dioxide (TiO_2) has the highest refractive index of any material commonly used in papermaking, and as a result generates the highest light-scattering coefficient. It is found commonly throughout the world, but at relatively low levels of purity. Most often it is naturally found in beach sands.

There are two crystalline forms widely used in papermaking applications, anatase and rutile. The major difference between the two is crystal uniformity and size, which yields a slightly higher index of refraction in the rutile crystal (Fig. 6.8). There are two chemical processing manufacturing routes commercially viable for titanium dioxide production, the sulphate and the chlorine process. The older sulphate method may be used to produce both anatase and rutile, while the more recent chlorine process is utilised only for rutile crystal production.

While global consumption of titanium dioxide has decreased as alkaline papermaking is growing in volume, there are still approximately 100,000 tonnes used annually for filler applications in printing and writing grades. The majority of that demand stems from North America, where opacity in both coated and uncoated freesheet grades is of paramount importance. However, the largest industrial application for titanium dioxide fillers is in laminate grades for flooring and furniture. These laminates comprise layers of resin impregnated fibres, which have a dense structure with little void volume. Thus, titanium dioxide can be used

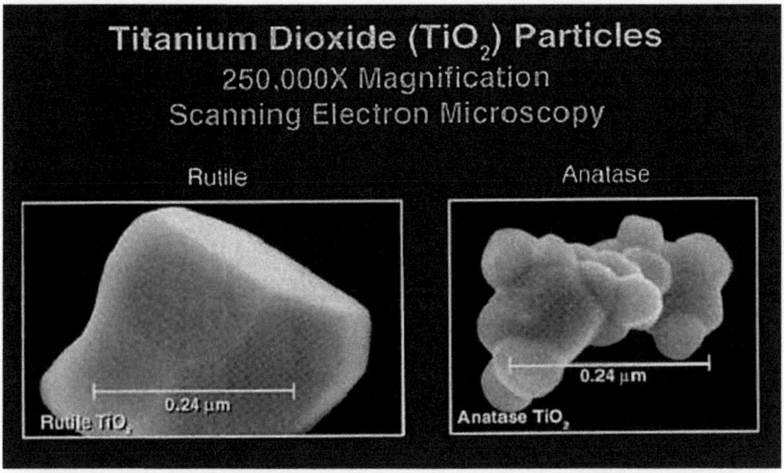

Fig. 6.8 Anatase and rutile titanium dioxide

in subsequent layers to mask the wood and create an aesthetically pleasing finish. In certain white laminates titanium dioxide levels can approach 50%. Approximately 250,000 tonnes are consumed in these speciality paper/paperboard composite applications annually.

6.2.4 Talc

Talc is a hydrated magnesium sheet silicate with the chemical formula $Mg_3Si_4O_{10}(OH)_2$, comprising 31.9% MgO, 63.4% SiO_2 and 4.7% H_20. Most papermaking products exist as platy crystals of alternating MgO/ MgOH octahedral layers between two SiO tetrahedral layers. The silicate layers are held together by way of weak Van der Waal's forces and are easily separated into plates of varying sizes (Fig. 6.9). Talc is naturally hydrophobic or organophilic which contributes to its functionality as a print enhancing pigment.

Talc may be added to the wet-end of a paper machine as either a pitch control agent or as filler for specific printability characteristics. By volume, the largest use for talc in paper is pitch passivation (where 0.5–2% of the furnish may be used to detackify pitch and/or stickies); however, an estimated 700,000 tonnes per year is consumed globally for paper filler applications. Asia, followed by Europe, is the largest consumer of talc as filler, where it is mostly used for rotogravure printability enhancement. In Europe, talc is most often used as filler in wood-containing grades, primarily supercalendered, as a co-pigment with either kaolin or carbonate. Rotogravure printability, in particular, is significantly improved (Lasmarias, 2002). In India, south korea and China, talc is often used as a filler in uncoated wood-free grades due to its wide availability and low cost.

Fig. 6.9 Platy talc

While talc is a relatively common mineral, deposits of high purity are rare. Commercial deposits are found in western USA, Finland, China, India and Western Europe. Talc is the softest mineral, exhibiting a hardness factor of 1 on the Moh's scale, and delivers a smooth tactile characteristic to the final sheet when used at the appropriate level. However, in China and India, local mining methods may lead to a high content of abrasive contaminants which may negatively impact wire life on a paper machine.

Talc is most often delivered in dry form to the papermaker and pigment makedown occurs using a small slurry unit on site. Due to the hydrophobic nature of talc, a wetting agent must be used as the first step in slurry makedown, often followed by a traditional polyphosphate dispersing agent.

6.2.5 Amorphous Silicates

Highly structured, agglomerated amorphous silicates are a class of synthetically prepared pigments that are used in a variety of paper grades, where brightness, opacity, bulk, coefficient of friction and ink receptivity are considered critical properties. In many cases, precipitated silicas serve as partial or complete replacements for titanium dioxide.

The manufacture of amorphous silicas is generally a proprietary process, which uses sodium silicate as a starting raw material. By controlling the reaction conditions, the preferential production of pigmentary silicas rather than silica sols may be

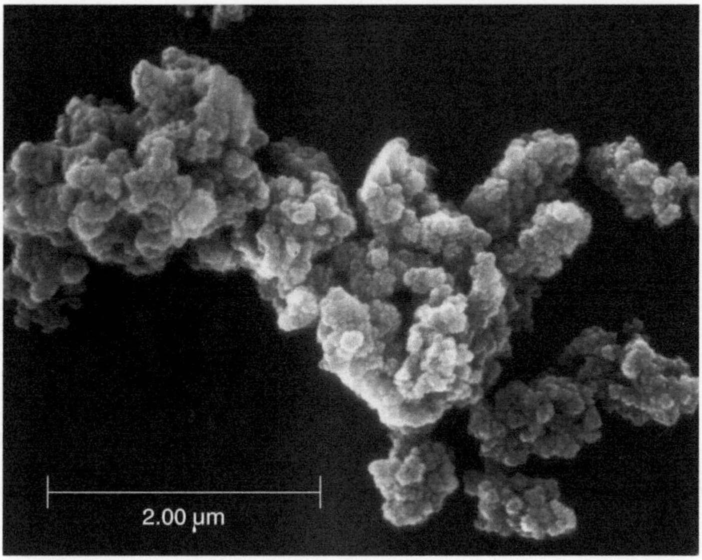

Fig. 6.10 Amorphous silicate

achieved. These highly ordered agglomerated aggregates exhibit a large degree of solid/air interfaces which contribute to the product's light-scattering characteristics as seen in Fig. 6.10 (Alderfer and Crawford, 1997). It is estimated that 30,000 dry metric tonnes of precipitated silica is used for paper filling applications, primarily in North America and Europe.

6.3 Fundamental Filler Characteristics

It is clearly evident that numerous mineralogies are utilised for paper filling applications; however, these minerals are all classified in much the same way. Particle size and size distribution, pigment brightness, refractive index, particle shape, and specific surface area are quantifiable characteristics that can be used to predict how the pigment will perform in various paper applications. To a lesser degree, particle charge, or zeta potential, plays a role in how the filler interacts with various paper chemical additives and influences the manner in which the mineral is retained in the paper web.

A generalized table of fundamental filler characteristics is given in Table 6.1 for a range of carbonate and kaolin grades as well as typical filler titanium, talc and precipitated silica. By developing an understanding of how these fundamental

Table 6.1 Typical characteristics of common fillers

Mineralogy	ISO brightness	Average particle size (μ)	Particle size distribution	Specific surface area (m^2/g)	Particle morphology
European kaolin filler	78–82	1.5–2.5	Broad	6–10	Plates relatively high aspect ratio
Engineered European kaolin	82–84	1.5–2.0	Somewhat steeper	10–12	Plates high aspect ratio
North American kaolin filler	80–85	0.3–1.5	Broad	15–20	Blocky stacks of plates, low aspect ratio
Brazilian kaolin filler	88–90	1.5–2.5	Broad	4–10	Blocky plates, low aspect ratio
Delaminated Kaolin	86.5–89	1.0–1.5	Broad	10–14	Plates high aspect ratio
Calcined Kaolin	90–93	0.4–0.7	Steep/ narrow	15–20	Aggregate, low aspect ratio
Coarse ground calcium carbonate	94–97	1.5–2.0	Broad	5–8	Blocky; low aspect ratio
Fine ground calcium carbonate	95–97	0.7-.0	Broad	10–13	Blocky, low aspect ratio
Engineered ground calcium carbonate	95–97	0.7–2.0	Somewhat steeper	5–13	Blocky, low aspect ratio
Rhombohedral precipitated calcium carbonate	95–98	1.0–3.0	Steep/ narrow	4–10	Blocky, low aspect ratio
Scalenohedral precipitated calcium carbonate	95–98	1.0–3.5	Steep/ narrow	5–14	Clustered rosette aggregates, low aspect ratio
Aragonitic precipitated calcium carbonate	95–98	0.75–2.5	Steep/ narrow	6–18	Clustered needle aggregates; low aspect ratio
Titanium dioxide	>97	0.2–0.4	Steep/ narrow	5–7	Spherical, low aspect ratio
Talc	80–85	1–8	Broad	3–6	Platy high aspect ratio
Amorphous Silica	>98	0.6–0.8	Steep/ narrow	40–80	Aggregate, low aspect ratio

characteristics combine to yield functionality in the papermaking process, a papermaker is better equipped to maximise performance while minimising cost.

6.3.1 Pigment Brightness

Pigment brightness typically ranges from the high 70's to the high 90s (ISO), and is affected by impurities which alter the light absorption characteristics of the mineral. In the case of hydrous kaolins magnetic separation, ozonation, bleaching and flotation may be utilised to remove impurities and thus improve pigment brightness. In this manner, hydrous kaolin brightness can be increased from the low 70s to as high as 92 (ISO). Fully calcined kaolins generally exhibit brightness in the range of 92–93.

Carbonates are higher in brightness by nature, and range from 94 to 98 (ISO) depending on the processing and mineralogical impurities inherent in the limestone. GCC processing often utilises flotation to improve brightness with a final range of 94–97, while PCC generally ranges from 96 to 98. However, commercially available chalk grades often have brightness values lower than 90, depending on the impurities found in the specific deposits.

Talc brightness usually ranges in the mid 80s and is not typically controlled through specific processing. Finally, titanium dioxide and precipitated silica generally exhibit ISO brightness values greater than 97.

While pigment brightness does contribute to sheet brightness, opacity, and shade development, it does not significantly impact light scatter. This is important to understand since brighter pigments of similar shapes and sizes usually means lower light absorption. Opacity suffers as the light absorption of the pigment decreases even though this can mean that the brightness of the sheet increases and the shade shifts to the blue. To get both higher opacity and brightness at the same time the light scatter of the sheet must be increased as opposed to simply increasing the absorption coefficient.

6.3.2 *Refractive Index*

Refractive index is an intrinsic property specific to each mineral type and is not altered by most processing. The index of refraction is defined as the ratio of the speed of light travelling through a vacuum to the speed of light travelling through the given particle. As light travels from one refractive environment to another, the angle of the light path is modified and the overall light scattering and absorption characteristics of the media are affected. The degree of modification is dependent upon the difference in refractive index (RI) at the interface.

A list of refractive indices of common papermaking materials is given in Table 6.2 (Lide, 2002). Of significant interest, titanium dioxide exhibits a much higher RI than any other papermaking component, and it should come as no surprise that use of titanium dioxide as a filler results in sheets with very high opacity and brightness. However, while many of the common papermaking materials exhibit similar RI, the paper producer may capitalize on particle packing to alter the refractive index of the sheet as a whole. For example, by incorporating materials of varying refractive index (fibre, air and mineral) and engineering a sheet of higher bulk, the angle of refracted light will change and improved light-scattering efficiency may be achieved.

Figure 6.11 demonstrates this concept by illustrating the relationship between bulk and light scatter in highly filled wood-containing grades. Use of platy kaolins alone results in sheets of lower bulk, and relatively low light scatter. As pigments with varied dimensions (as well as shape and size) are incorporated into the sheet,

Table 6.2 Refractive indices of common papermaking materials

Papermaking material	Refractive index
Anatase TiO_2	2.49–2.56
Rutile TiO_2	2.61–2.90
PCC	1.49–1.65
GCC	1.56
Kaolin	1.55–1.57
Calcined kaolin	1.56
Amorphous silica	1.45
Aluminum trihydrate	1.57
Cellulose	1.55
Starch	1.49
Air	1.00

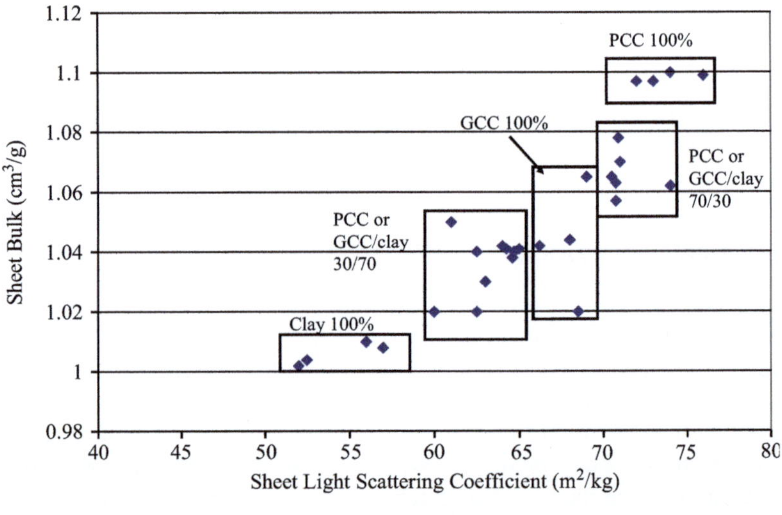

Fig. 6.11 Altering the refractive index of the sheet through use of filler can result in improved light-scattering efficiencies

overall bulk increases along with a corresponding increase in light-scattering coefficient. As the varied shape pigments separate wood fibres, the overall void volume is increased and light scattering is positively impacted through a greater index of refraction of the composite paper media.

6.3.3 *Particle Morphology*

Common fillers exhibit a wide array of particle shapes that are either naturally occurring or specifically synthesized. Particle morphology can be divided into two broad classes: discrete or aggregated. Fillers exhibiting a discrete morphology consist of a population of singular crystals with varying uniformity, while aggregated fillers are typically engineered crystals resulting in clustered particles of similar size and shape within the population.

6.3.3.1 Discrete Particles

Hydrous kaolins, standard GCC, rhombohedral PCC, talc and titanium dioxide may be considered products that comprise discrete particles. These filler products typically range from plates to spheres to blocks of varying uniformity.

Talc and hydrous kaolins are naturally platy by nature, and degree of platiness can be enhanced by the delamination process. Oftentimes, these products are differentiated by aspect ratio which is defined as the ratio of plate diameter to plate thickness. Key benefits of shape-engineered platy particles include reduced sheet permeability along with improved smoothness and printability.

Historically, pigments of very high aspect ratio were not commercially utilised based on poor rheology and handling characteristics of high solids slurries. However, modifications to the particle size distribution, as well as technological advances in processing have allowed a reduction in handling concerns while delivering products of increased functionality to the market.

For example, coarse kaolins of high aspect ratio are considered excellent choices for supercalendered grades based on ease of retention and improved print quality in the final sheet. As shown in Fig. 6.12, replacement of only 25% of the PCC with platy kaolins can significantly impact sheet permeability. In this example, five

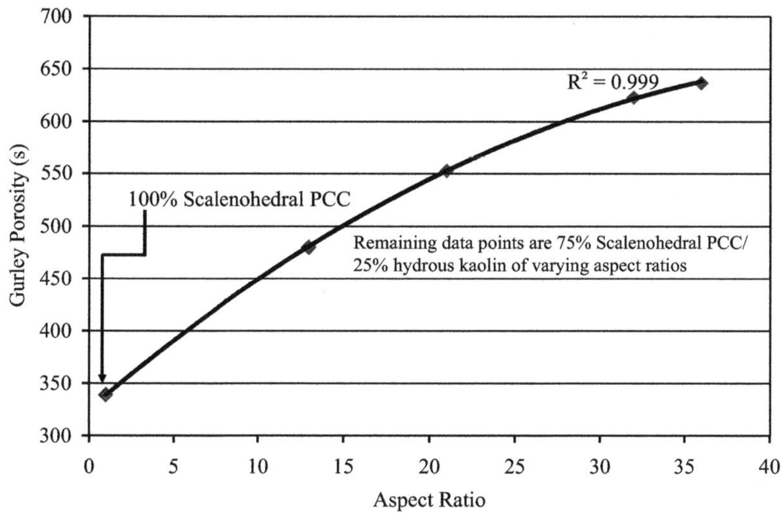

Fig. 6.12 Impact of aspect ratio on SC sheet permeability (TMP/GWD/Kraft furnish; 52 gsm, 28% filler loading)

pigment systems were evaluated at 28% loading in a TMP/GWD/Kraft furnish. Four of the pigment systems were 75% scalenohedral PCC with 25% kaolin of varying aspect ratios (shown along the *x*-axis). The control system consisted of 100% scalenohedral PCC and is denoted as the data point exhibiting an aspect ratio of ~1. The data indicate a significant inverse relationship between sheet permeability and aspect ratio; i.e. as the aspect ratio (or platiness) of the kaolin component is increased, sheet permeability decreases dramatically. This can help reduce ink set rate and improve ink mileage in offset print applications where use of PCC as sole filler has been a challenge (McLain, 2006).

GCC and rhombohedral PCC exhibit a low aspect ratio and are blocky by nature. These products offer reasonable drainage characteristics on machine and tend to open the surface more substantially than platy particles.

Titanium dioxide is relatively spherical by nature, and is sized appropriately for superior light diffraction, which will be discussed at greater length in the section reviewing particle size considerations.

6.3.3.2 Aggregated Fillers

Calcined kaolin, scalenohedral PCC, aragonitic PCC and amorphous silicates are examples of aggregated fillers, i.e. these products consist of crystals which are synthetically clustered into repetitive shapes of similar size throughout the population.

In the case of calcined kaolin, individual kaolin plates are thermally fused into a clustered shape. With scalenohedral and aragonitic PCC, either a rosette or repetitive needle structure is formed through carefully controlled reactions of milk of lime and carbon dioxide. Amorphous silica products are chemically aggregated into repeating silica rings.

In each of these product types, air is used as an integral component within the structure, thereby creating relatively porous fillers with varying degrees of void volume. These air voids are key to the performance of the product, as they provide a number of benefits.

First, the air voids serve as a differing refractive environment relative to the surrounding mineral faces within the filler itself. This increases the effective refractive index of the aggregated filler and yields a higher light scattering of the product, as seen in Fig. 6.13 (Gate and Husband, 1986).

Second, these voids tend to yield fillers of higher bulk density which in turn provides higher bulk to the finished sheet compared to fillers exhibiting discrete morphologies. As shown earlier, higher sheet bulk results in greater light-scattering efficiencies, again through the mechanism of increased refractive index of the media.

Finally, fillers with high void volumes may serve as a carrier for certain colloidal substances in wood-containing systems. A good example of this is the attachment of triglycerides to porous fillers which can result in improved coefficient of friction and reduction of crêpe wrinkles in newsprint grades.

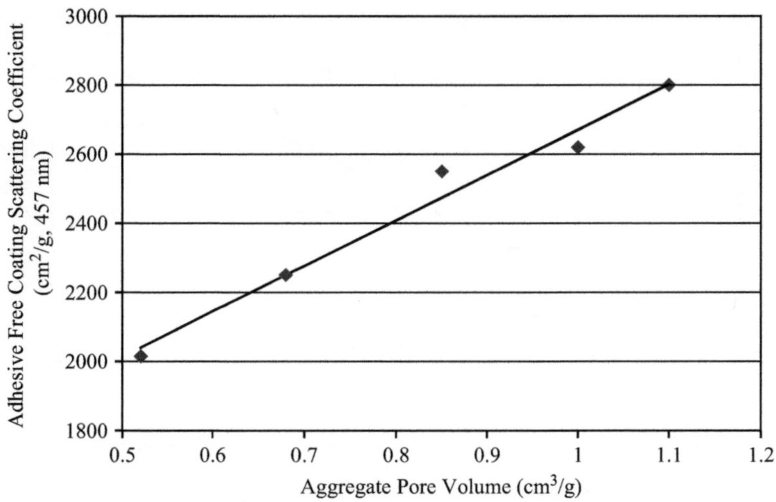

Fig. 6.13 Impact of internal void volume on light-scattering efficiency

6.3.4 Particle Size and Size Distribution

Particle size is generally measured by sedigraph, which calculates a particle's size based on relative settling rate. Though this method assumes a spherical geometry (and very few papermaking minerals are spherical in nature), it is the most commonly used method of particle size measurement within the pigment industry.

A typical particle size distribution curve is given in Fig. 6.14. When mass of particles is plotted versus equivalent spherical diameter on a logarithmic axis, the distribution is relatively Gaussian around a mean particle size. The steepness of the particle size distribution is described by the standard deviation of the distribution. A more typical depiction of the curve is to plot the percentage of particles versus equivalent spherical diameter. For simplicity, steepness can be defined as the ratio of two population points within the distribution. In this case, the steepness factor is defined as:

Steepness = 100× (d30/d70)

The higher the steepness factor, the lower the deviation from the mean particle size. Based on processing technologies, hydrous kaolins and GCCs have a significantly broader distribution than synthetic fillers like calcined kaolins PCC, and titanium dioxide. Removing a portion of the very large particles, the very small particles, or both will steepen the particle size distribution. In this way, the particle size distribution can be altered to improve light-scattering efficiency. Figure 6.15 shows the impact of particle size on light scatter. In the case of these scalenohedral PCC crystals, light scatter peaks at a particle size of 1.5–1.6 μm. As particle size increases, there is a significant drop in light-scattering efficiency. It should be noted that optimal particle size of filler is system-dependent and will vary among paper machines.

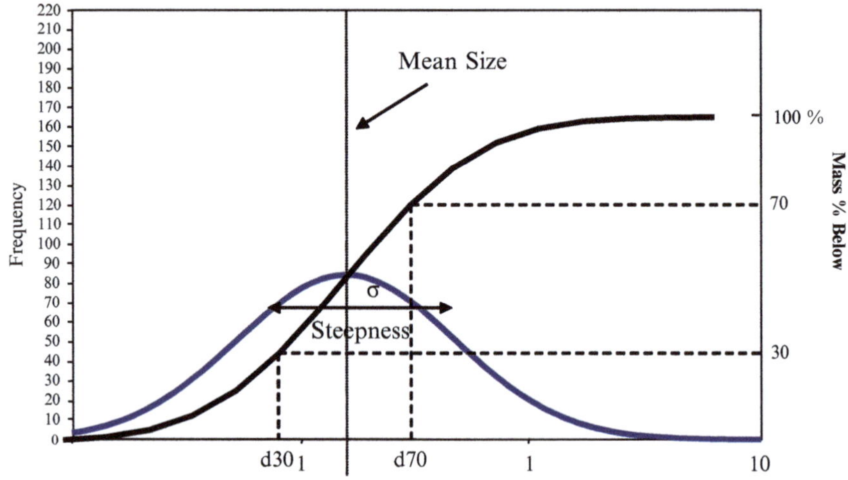

Fig. 6.14 Particle size distribution as measured by sedigraph

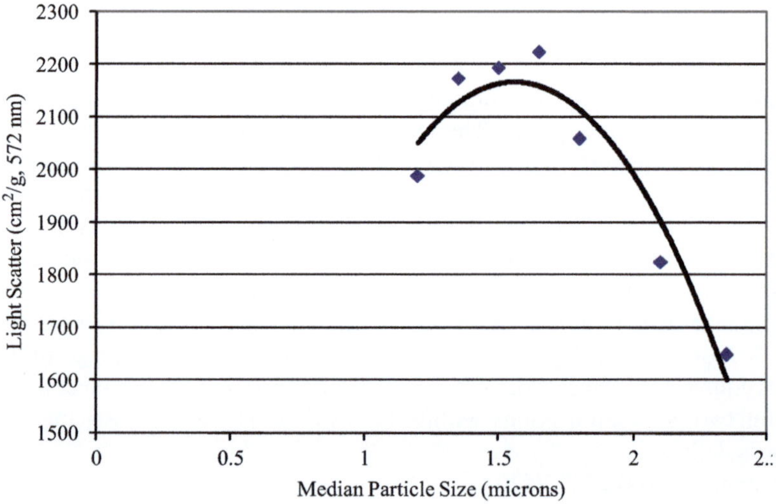

Fig. 6.15 Impact of scalenohedral PCC particle size on light scatter, 8% loading, 65 gsm freesheet

Similarly, Fig. 6.16 depicts the impact of kaolin particle size and steepness on light-scatter development. As expected, smaller particles with steeper distributions yield a higher light-scattering coefficient. There is a threshold particle size, beyond which narrowing the particle size distribution does not improve light scatter, as demonstrated by the 4 μm products. This is due to the very poor degree of light diffraction around particles of larger dimensions.

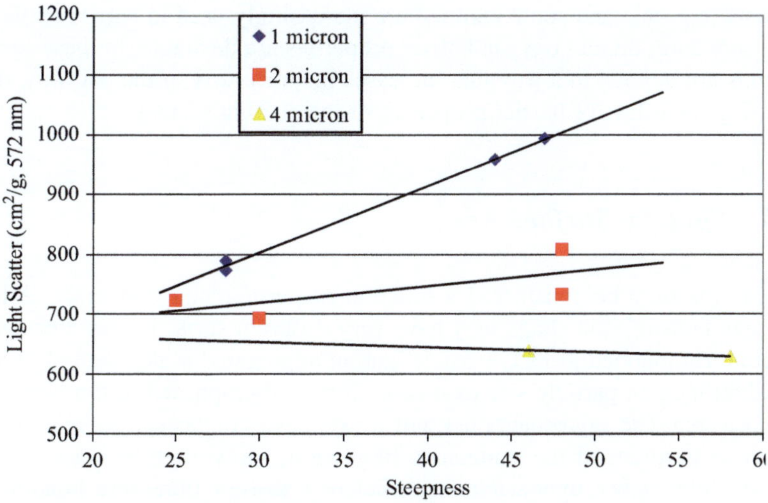

Fig. 6.16 Impact of particle size and steepness on light scatter

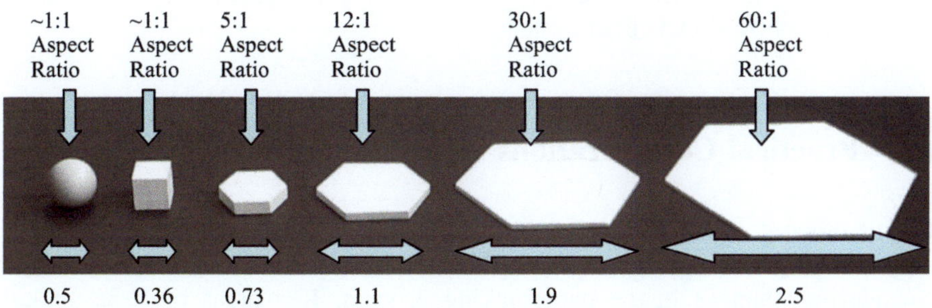

Fig. 6.17 These models are to scale of common papermaking materials that all measure as 0.5 µ equivalent spherical diameter by sedigraph; as expected, they each have very different performance characteristics

Finally, it is important to understand that minerals may appear similar in particle size as measured by sedigraph (or even optical light-scatter devices), but differ significantly in particle shape. As seen in Fig. 6.17 pigments which measure equivalent spherical diameter by sedigraph can have very different shape factors, or aspect ratios. One might expect these pigments to behave very differently with respect to final sheet property development, and indeed, that is the case. For example, the ultra-fine glossing clay is fine and blocky with a very low shape factor (5:1). Kaolin of this type is very efficient in gloss development as it deforms well under the calender. Conversely, the ultra-fine platy has a relatively high shape factor (30:1) and is quite efficient at reducing air permeability and enhancing printability characteristics as it tends to align lengthwise in the machine direction. Hyperplaty pigments (e.g. talc and some kaolins) are increasingly used in barrier applications where moisture, oil and oxygen barrier properties are desirable. In these cases, the pigments are aligned in a way that increases the tortuosity of the pathway, thereby increasing the effective barrier properties of the coating system.

6.3.5 Specific Surface Area

Surface area may be considered a function of particle size and shape. Fillers of same size but different shape may have very different surface areas. For example, a discrete rhombohedral PCC particle and an aggregated scalenohedral PCC with equivalent median particle size as measured by sedigraph, will differ significantly in surface area. The aggregated filler will always have a higher surface area compared to discrete particles of the same size. In general, fillers with higher surface area tend to yield higher light-scattering efficiency, disrupt fibre/fibre bonding more significantly and exhibit a higher chemical demand than fillers with lower surface area. Thus, a freesheet producer can expect to achieve high opacity and brightness through the use of an aragonite or scalenohedral PCC, but would also expect a weaker sheet with higher demand for sizing chemical compared to a similar particle sized rhombohedral PCC filler system.

6.4 Practical Considerations When Using Filler

Use of mineral will impact paper machine runnability in a number of ways, and understanding how specific types of fillers interact with fibre is important when considering changes in the papermaking process. Key areas of focus include interaction with light, pigment retention, impact on drainage and drying rates, point of addition and possible interactions with other papermaking components.

6.4.1 Filler Interactions with Light

As one of the primary reasons for utilising filler is the development of optical properties, it is necessary to establish a cursory understanding of the interaction of various filler types with light of varying wavelengths.

The interaction of light with filled papers is usually described in terms of Kubelka-Munk theory. This is a macroscopic, one dimensional, continuum theory that describes interactions between light and media in terms of scatter and absorption. The theory has recently been improved with a correction related to the mean optical path lengths before scattering or absorption (Yang and Kruse, 2004; Yang, 2004). The correction only applies when scattering is greater than absorption, which is usually the case with printing and writing papers. Although this relatively simplistic theory can generally be used to describe brightness and opacity, or reflectance and transmittance at different wavelengths, it does not address specular reflectivity, also known as gloss.

Absorption of light is an intrinsic property that depends upon the chemical species that are present in a media. Different species will absorb light at different wavelengths, resulting in colouration of the media. The absolute amount of light that is absorbed also depends upon the optical scatter. If a photon cannot easily pass through the media, but is scattered many times, it will have a longer optical path length within the media, and will have greater opportunity for being absorbed.

Optical scattering in paper is an extrinsic property that depends on how the paper has been produced. For example, a sample of paper can be calendered differently and will have different optical scatter characteristics. Thus, it is not possible to quote an absolute scatter value for a particular type of paper fibre or mineral filler pigment. It is only possible to quote a value for the system of fibre and filler under a very particular set of production parameters, and to track how the value changes as the system and production parameters are changed.

Optical scatter is a composite of several different physical phenomena, including reflection and refraction at material interfaces as well as diffraction by small particles. Reflection and refraction phenomena are dependent upon the refractive index of a material. A material with a high refractive index, like TiO_2, may thus impart high optical scatter to a paper system. Composite pigments, including calcined kaolin and scalenohedral PCC, for example, provide internal air-to-mineral interfaces to increase the opportunities for reflection and refraction. In addition to the index of refraction, diffraction is strongly dependent upon particle size, with the strongest interactions occurring for particles similar in size to optical wavelengths or a fraction thereof.

Mineral fillers affect the optical properties of paper primarily by increasing the optical scatter, along with some absorption. The high optical scatter achievable with mineral fillers is primarily related to the small size of the fillers in comparison to paper fibre. The small particle size translates to a large number of particles, providing many material interfaces for reflection and refraction, as well as a possibility for non-trivial diffraction. Another way that fillers increase scatter is by decreasing the contact between fibres, again providing more surface area for reflection and refraction, but decreasing fibre-to-fibre bonding. One final way in which fillers may influence paper optics is to increase gloss. This may occur due to the pigments filling in space between fibres at the paper surface, thus presenting a smoother, more mirror-like surface for the light to reflect from (McLain and Wygant, 2005).

6.4.2 Retention

Particle size and surface area play a role in chemical demand required to sufficiently retain pigments within the web. Likewise, dispersants often associated with high solids filler slurries tend to interfere with many conventional retention aid chemistries and should be taken into consideration when grade development strategies are formed. Particle charge, or zeta potential, is often an indicator of the relative ease of retention for many fillers.

6.4.3 Drainage and Drying

Extension of fibre with filler almost always results in reduced drying demands, as minerals do not absorb water compared to fibre. However, aggregated fillers with very high surface area and internal void volume may result in a wetter web going to the press section. Figure 6.18 shows a comparison between scalenohedral PCC and GCC filler in press solids. Discrete particles, like GCC, typically drain more quickly compared to similarly sized aggregated particles, like scalenohedral PCC.

Energy conservation is an area where cost savings may be achieved in most paper mills. By carefully considering filler type and level, some ease of drying may be gained. If strength considerations can be balanced, there may be some potential to increase productivity at the same drying energy, which could result in significant gains in profitability.

6.4.4 Point of Addition

With the exception of calcium carbonate, mineral fillers are inert across the papermaking pH range. Neither PCC nor GCC may be used in systems exhibiting a pH lower than 6.5, and special attention must be paid to systems in the neutral papermaking range (6.5–7.5) to ensure that the carbonate filler does not dissolve and create both foaming and calcium deposit issues.

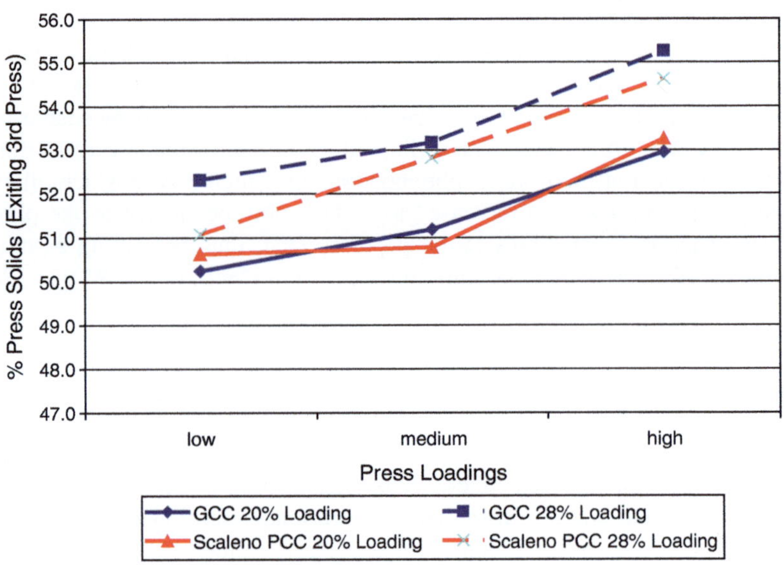

Fig. 6.18 Impact of particle shape on dewatering

Mineral fillers should be added at points which promote good mixing. A typical thick stock addition point is the suction side of the machine chest pump. Addition of filler in thick stock will give increased residence time and may improve ash retention nominally depending on the addition points and residence time of the retention chemistry. A typical thin stock addition point is the primary or secondary fan pump, which yields quick response to elevated levels of minerals, which is useful for mills that make several significant grade changes.

6.4.5 Potential Interactions with Other Papermaking Components

Probably the most cited instance of mineral interaction with papermaking materials is the quenching effect titanium dioxide has on optical brightening, or fluorescent whitening agents (OBA or FWA). These products function by absorbing near UV light energy and re-emitting in the visible blue spectrum. Titanium dioxide absorbs the same light energy and therefore competes with the OBA in a negative manner. Thus, it is important to separate OBA/FWA and titanium dioxide addition points to minimise this interaction. Another option is to use anatase titanium dioxide, which does not absorb the near UV radiation as readily as the rutile titanium crystal. Finally, in coated grades, Kwoka and Logan (1994) show that splitting the titanium usage between the base stock and coating yields the most efficiency.

Regarding other fillers, it is usually best to avoid close addition of mineral with materials that have a floccing effect (e.g. starch or coagulants), as these may agglomerate the filler particles and result in lowered light-scatter efficiency due to crowding or inefficient particle packing. However, there are cases when starch or coagulants may be used to increase ash content by improving retention and agglomerating the filler so there is less impact on overall sheet strength.

In summary, it is of paramount importance for the papermaker to have a clear idea of the performance attributes which are to be achieved by filler. With that information and a cursory understanding of how the fundamental filler characteristics scale up to application performance, cost-effective paper grade design is possible.

References

Lasmarias V, et al (2002) Productivity and Quality Enhancement of SC Papers with Talc. Paper presented at Paptec Conference, Montreal, Quebec, February 2002

Alderfer G, Crawford R (1997) High-Structure Amorphous Silica Pigments in Paper. In: Hagemeyer R (ed.) Pigments for Paper, TAPPI Press, Atlanta, Georgia

Lide D (ed.) (2002) Handbook of Chemistry and Physics, 83rd Edition, CRC Press, London

McLain L (2006) Multi-Pigment Platforms: Maximizing Value through Functional Grade Design. Paper presented at IntertechPira's Pigments and Fillers for Applications Conference, Atlanta, GA, 28–29 September 2006

Gate LF, Husband JC, IMERYS internal research report, unpublished work, 1986

Yang L, Kruse B (2004) Revised Kubelka-Munk Theory. I. Theory and Application. *Journal Opt. Soc. Am. A* 21(10): 1933

Yang L, et al (2004) Revised Kubelka-Munk Theory. II. Unified framework for homogeneous and inhomogeneous optical media. *Journal Opt. Soc. Am. A* 21(10): 1942

McLain L, Wygant R (2005) Fundamental Filler Characteristics for Maximum Sheet Performance, Paper presented at AEL Metsko's Stock Preparation and Uses of Various Pulps Workshop, Munich, Germany, November 2005

Kwoka R, Logan T (1994) Titanium Dioxide Performance Comparison Between Coated Paper Basesheets and Coatings. *TAPPI* 77(10): 136–144

Chapter 7
The Improvement of Dry Strength by Synthetic Polymers

Stan Jenkins

7.1 Introduction

In the 1950s a new family of papermaking additives was developed – polyacrylamides [1]. Polyacrylamides are made by polymerising acrylamide solution using free-radical catalysts, to form a linear, non-ionic polymer [2].

The average molecular weight can vary from a few thousands to several millions, depending upon the polymerisation conditions, and may be formed as a powder, water-based solution or water in oil emulsion.

Since the original introduction of polyacrylamides, considerable development has taken place, and their use is now regarded as standard in most mills. Three major applications have become established: retention, water treatment (from save-all to effluent plant) and as a dry strength resin.

Non-ionic polyacrylamides, although somewhat effective as retention aids, are essentially non-substantive to fibres. Anionic groups can, however, be formed on the polyacrylamide by introducing sodium acrylate or acrylic acid to the monomer solution to form a copolymer. The quantity of sodium acrylate will define just how anionic the polymer becomes. Typical substitution levels are in the range of 5–10 mole%, although higher levels are possible. Lower charge anionic polymers can be produced by hydrolysing some of the acrylamide, thus avoiding the need to use two different monomers.

Relatively low molecular weight (around 250,000–1,000,000) versions of this polymer became the first commercial synthetic dry strength resins in 1955 [3, 4]. This product, when used in the presence of alum, was able to offer improvements in tensile strength of up to 25% in a range of papers, including printing papers, hygiene papers and packaging papers and board. Since that time, many new resins have been developed, both anionic and cationic, although variants of this original product are still being extensively used in mills today.

In recent years, there has been a growth in the interest in synthetic dry strength agents and new approaches have been developed, including the use of glyoxylated polyacrylamides and polyvinylamine chemistry. In some cases these polymeric

S. Jenkins
Papertex Ltd., P.O. Box 11, Skipton, BD23 9AF, UK

I. Thorn and C.O. Au (eds.), *Applications of Wet-End Paper Chemistry*,
DOI 10.1007/978-1-4020-6038-0_7, © Springer Science+Business Media B.V. 2009

systems are used in addition to the use of natural products such as starch, in order to boost the strength beyond the level normally seen with these materials.

7.2 Benefits of Dry Strength Resins

There are many benefits to be obtained from using dry strength resins. The extent of these benefits does however depend on the inherent strength already in the sheet. It is very difficult to increase the strength of an already very strong sheet. It will also be extremely difficult to increase the sheet strength above the natural strength of the papermaking fibres, as this is normally the weakest point of the sheet.

The benefits of using a dry strength additive can be split into two main groups [5], those that reduce costs, and those that improve the quality of the paper. Additive costs can be reduced through the replacement of existing materials by the resin. This can include fibre substitution by lower cost materials such as secondary fibre or filler, reduced basis weights, or the promotional effects of the resin on other additives such as sizes, wet strength resins, or dyestuffs. Reduced operating costs can be obtained through reductions in refining, which can lead to improved dewatering, easier drying, higher machine speeds, easier re-pulping, and lower energy costs. Many synthetic dry strength agents will also give retention and drainage enhancements, and in some cases have been able to replace an existing retention aid programme as well as providing improved dry strength.

Quality benefits can be achieved from such advantages as increased bulk, higher strength, greater opacity, enhanced printability, improved absorbency or superior tissue softness. Collateral benefits can also be achieved, such as higher retentions, increased sheet uniformity, and improved machine runnability. Depending upon the nature of the resin and the furnish, porosity gains may also be seen.

7.3 Strength Additives Versus Refining

The most common way of increasing the dry strength of a paper or board, is through fibre choice, refining or beating. Refining applies energy to the papermaking fibres and fibrillates them. This creates tiny fibrils attached to the side of the fibre and releases hemicellulose. Hemicellulose is composed of short-chain polymers of various types of sugar, and has been found to contribute significantly to the dry strength in the sheet [6]. The extent of the fibrillation depends on the pulp type, as well as the type of refiner. As the degree of refining increases, the pulp will retain more and more water (becoming 'wetter beaten'), and the fibres will conform more and more to each other. This results in a denser sheet, which in turn reduces the sheet bulk.

This more compact sheet leads to lower porosities and reduced opacity. The ultimate version of this is natural tracing paper, where the chemical pulp is so heavily beaten, that the sheet becomes virtually transparent. This also has the effect of imparting a high level of dry strength to the sheet.

Farley [7] reported the changes in handsheet properties that occur when a bleached softwood Kraft pulp is 'refined' in a laboratory beater. When a lightly beaten pulp was taken as the standard, increases in internal bond (Scott bond or *z*-direction strength), fold, burst and tensile strengths were seen. These were accompanied by losses in calliper and opacity. Farley also showed that variations in press loadings will have similar effects to refining. A denser more highly bonded sheet is formed, increasing dry strength (except tear) and reducing calliper and opacity. The effect of refiners on shortening the fibre length will have an adverse effect on tear strength.

From this we can see that the papermaker is locked into a certain sheet density for a particular dry strength. Variations in furnish can help, but often these are accompanied by other effects such as reducing opacity or increasing cost. Other compromises may also need to be made in terms of stiffness, softness (tissue), opacity, porosity, tear or drainage.

In order to increase dry strength without these compromises, dry strength additives are used. These give the extra strength required, but without the loss of calliper or opacity. In many cases, the papermaker is actually seeking other benefits that would normally result in a loss of strength. The use of a dry strength additive allows this to be achieved.

Examples of this include the tissue maker who wants higher softness, without a loss in strength. Although changes in the tissue bulk, through furnish changes, pressing, crêping or refining will give the desired softness, these inevitably result in a loss of tensile strength, or significantly higher costs.

Dry strength resins are widely used in this application to recover the loss in strength produced in developing a bulkier softer sheet. Similarly, mills wishing to substitute cheaper fibre, including secondary fibre, or wanting to reduce basis weight, will often use a dry strength resin to maintain the needed strength characteristics of the paper or board.

Table 7.1 shows the trade-offs that can be made with different methods of achieving increased dry strength. The papermaker must assess his needs versus the disadvantages of alternative approaches. The use of a chemical system can offer other collateral benefits as well as disadvantages. Examples are improved retention and drainage, increased porosity, promotion or interference with other chemical additives, and increased BOD and COD in the water system.

7.4 Types of Synthetic Resin

Today, the most common types of synthetic dry strength additive are based on polyacrylamide although polyvinylamine strength additives are growing in popularity. This is because of the relatively low cost, and the ability to readily put an ionic charge onto these polymers. Other resins such as polyvinyl alcohol or lattices are used, but generally these are seen in surface applications to the paper, rather than as wet-end additives. Similarly styrene acrylic resins have shown superior performance when applied through surface application rather than as stock additives [8].

Table 7.1 General effects of refining and dry strength resins

	Refining	Dry strength resin
Fibre		
Fibre bonding	Increase	Increase
Fibre length	Decrease	None
Fibre specific surface area	Increase	None
Fibre flexibility	Increase	None
Fibre primary wall removal	Increase	None
Sheet		
Dry strength	Increase	Increase
Tensile	Increase	Increase
Burst	Increase	Increase
Internal bond	Increase	Increase
Fold	Increase	Increase
Stiffness	Decrease	Minimal increase
Tear	Decrease	Decrease
Calliper	Decrease	None
Opacity	Decrease	None
Porosity	Decrease	None
Absorbency	Decrease	None
Sizing	Decrease or increase	Slight increase
Dimensional stability	Decrease	None
Formation	Improve	None
Process		
Retention	Increase or decrease	None or slight increase
Drainage	Decrease	None or slight increase
Dryer adhesion	Increase	Increase
Drying rate	Decrease	None
Power consumption	Increase	None
Stock		
Freeness	Increase	None
Fines content	Increase	None
Temperature	Increase	None

The early anionic polyacrylamide resins are still widely used, both in powder and liquid form. The natural anionic charge on cellulose fibres, and the contaminants commonly found in papermaking systems, led researchers to seek methods of putting a cationic charge on these chemicals, thus making them more substantive to fibres.

One such method is to form a 'Mannich' product of polyacrylamide which produces a cationic polymer under acid conditions. At the time these products were developed, rosin/alum papermaking was still generally used, so this was not a problem. As papermaking pHs rose, however, these products were generally favoured less.

Including cationic monomers such as diallyldimethylammonium chloride or dimethylammoniummethyl (meth)acrylate quaternized with methyl chloride (ADAME-Quat) in a polyacrylamide produces a polymer that holds its charge across a much broader pH band, including alkaline conditions.

Since the original development of cationic polyacrylamides, several cationic monomers have been developed which are now in regular use in all applications for polyacrylamides. The use of these materials in paper is sometimes restricted due to a lack of governmental approvals for use in papers coming into contact with food. This can restrict the type of monomer and the molar quantity allowed. Despite these limitations, a number of cationic resins based on new monomers are now commercially available.

The addition of a crosslinking agent, glyoxal, to this type of resin gives wet strength properties as well as dry strength [9]. This material forms a covalent bond with the hydroxyl group on cellulose [10], and is widely used in applications where wet strength is tolerated or desired, such as towelling. The wet strength is of a temporary nature, losing effectiveness during prolonged exposure to water. This enables re-pulping of broke or waste paper to be readily achieved with no special treatment, such as the use of caustic or high temperatures.

The unreacted amide groups on the molecule allow considerable increases in dry strength to be achieved. This increase is independent of the wet strength mechanism, and is effective across a broad pH range (4.5–8.0). The dry strength obtained is often greater than that achieved with conventional dry strength resins, polyvinyl alcohol, starches or gums [11].

The glyoxylated resin has a low cationic charge, and is less effective in systems where there are high levels of anionics (anionic trash), such as secondary fibre furnishes. Here the resin complexes with both soluble and insoluble materials, thus reducing the adsorption of the resin onto the fibres. This can be overcome through the addition of cationic promoters (see below), or through careful charge control using other chemical additives in the furnish, such as polyamide wet strength resins or cationic sizes. The use of glyoxylated polyacrylamides in addition to natural materials such as cationic starch is growing in popularity, as the glyoxal will also give some crosslinking to the starch. This has the benefit of offering a synergistic effect on dry strength.

Poly(N-vinylformamide) (PNVF) was found [16] to provide a degree of dry strength without any contribution to wet strength. When PNVF is hydrolysed, either partially or fully, a cationic charge is developed on the molecule as polyvinylamines are formed. These became commercially available to the paper industry in the early years of the twenty-first century. Polyvinylamines give slightly more dry strength than PNVF, as well as providing some permanent wet strength. Both cationic and anionic forms of polyvinylamines are now commercially available [17] and the use of these products is growing.

The optimum molecular weight for dry strength has been the subject of debate. Linke [12] suggested that the degree of hydrogen bonding of the polymer is the same for the same molar quantity of any molecular weight resin. The higher molecular weight material will, however, cause a higher degree of flocculation, which in turn will mask any strength development. This means that low molecular weight products are preferred, to reduce flocculation effects.

The lower molecular weight materials offer other benefits. The solution viscosity of these products is largely dependent on the molecular weight of the polymer. Reducing the chain length means that higher solution strengths can be obtained, allowing reduced freight and storage costs. An additional benefit is that the molar concentration of the polymer will increase at the same active content as measured by weight. This means that there are more molecules to hydrogen bond with the fibres, and using Linke's argument, this should increase dry strength performance.

7.5 Methods of Use

Dry strength resins are generally supplied either as a powder or as a water-based solution or emulsion. In all cases, they can be readily dispersed in water.

The powder form of resin is considerably easier to handle, when compared to the more commonly known polyacrylamide products used for retention or effluent treatment. The much lower molecular weights of these materials allow dissolution to be made at much higher active solid levels, including over 10% active. Both manual and automatic make-down systems are available to achieve a homogeneous solution of the resin. The important point is to ensure that each powder particle is fully wetted out, and can dissolve in the bulk of the water. A collection of particles, which together are wetted around the outside, will take much longer to dissolve, and can result in gels. The presence of these on a paper machine may lead to breaks or spots in the paper.

Automatic make-down systems are usually more consistent in ensuring a constant solids solution without gels. Variations in water pressure can however affect this, so the installation of filters is always advised before the paper machine. Since the type of gel that would be formed is soft, the pressures created in an in-line filter can force the gel through the mesh. A low-pressure filter, such as at the inlet to the stock-holding tank, is advised.

Liquid, ready-to-use products are much more common these days. These can be pumped directly from a container or bulk storage tank, to the paper machine. Further dilution through the use of in-line addition of fresh water (not clarified) is recommended to ensure thorough mixing of often quite viscous products with the papermaking fibres. Again, good filtration is recommended prior to stock addition. The low solids content of these liquid products, combined with the sometimes high output of modern board machines may involve quite high volumes of chemical products being used. Consequently consideration of handling practices should be given when considering these materials.

7.5.1 Addition Points

Opinions vary as to the optimum point to add a dry strength resin, and indeed this can differ from one paper machine to another. It will also to some extent depend upon the specific objectives for use. The most commonly used point of addition is to the thick stock, after the last refining has been carried out. Strazdins [13] reported a laboratory study on a European secondary fibre furnish, where the benefits of resin addition to the long fibre fraction were demonstrated. Good mixing is essential to obtain the best performance from the product, so addition to a well-stirred chest or the down leg of the thick stock level box is ideal. If the dilute resin is added to a pipeline, optimum mixing will be achieved if a multipoint injection system is used.

Care must be taken to ensure that competition with other additives does not take place. In those cases where addition after the last refiner is not possible, success can be achieved if all major refining is completed before addition, and only light refining is applied afterwards.

Ensuring the charge balance in the system is maintained is essential to obtaining the best from all functional additives, and dry strength resins are no exception. When anionic resins are used, it is common to have a strong cationic promoter to ensure that the resin is fixed onto the anionic fibres. In the case of acidic papermaking, this promoter is normally alum or polyaluminium chloride. Under neutral and alkaline conditions, a synthetic cationic additive is often used. If high levels of wet strength are also required, such as in towelling or label papers, then the wet strength resin performs the function of promoter as well as its more general purpose: similarly, cationic sizes, or their promoter, can do the same task. If no or insufficient cationic functional chemical is present, then a separate cationic polymer such as a polyamine is added. The addition of this material is usually prior to the dry strength additive. Although the promoter can go before refining, this process may release a lot of anionic materials such as hemicellulose, so addition after can give better results.

The use of such cationic promoters is not restricted to use with anionic resins. Benefits such as reduced resin usage or higher dry strength values have also been seen when used with cationic resins. An example of this occurred in a European tissue mill, which was using a dry strength resin to improve the softness of its toilet tissue. The addition of 1 kg/t of a cationic polyamine reduced resin usage by 40%, for the same physical results. Collateral benefits were increased drainage, improved runnability and increased felt life.

With the demand for greater performance from synthetic dry strength additives, this approach has been developed further, whereby two strength additives with opposite charge are used. Both materials give enhanced strength, but the combination often achieves greater strength than would be expected from the results when a single component is used. This approach has been used with starches, CMC, polyacrylamides, polyvinylamines and combinations of these components. Esser and colleagues [17] have reported on this approach giving very positive results at Europacarton Hoya.

When two dry strength components are added to achieve the synergy mentioned above, both components must be added after refining, but apart from each other. Typical addition points might be before and after the machine chest feed pump. This adds the resins to the thick stock, but ensures good mixing between each component. The further apart the two components, typically the better the results.

7.5.2 Dosages

The optimum dosage of a dry strength resin will vary from application to application, machine to machine, and grade to grade. Typical dosages generally fall in the range of 0.2–0.5%, based on active resin, although dosages as low as 0.05% have been successful. This is particularly so if a cationic promoter is used. Generally, performance starts to level off, when the dosage exceeds 0.7% active. Using a two-component or carrier system will result in higher total dosages being possible - sometimes exceeding 1% together.

The exact dosages may depend on the charge balance in the wet-end. The use of charge measurement systems, either laboratory-based or on-line, will help achieve the optimum dosage, especially when two component systems are in use. Judicious adjustment of the dosages of both additives based on the charge balance can achieve higher dry strength values.

7.5.3 Surface Addition

Dry strength resins have on occasion been used as a surface additive, with application at the size press. In a laboratory study an anionic dry strength resin as part of a size press formulation incorporating alkyl ketene dimer size and calcium carbonate, as well as the dilute resin, was applied to fluting paper to give a resin dosage of 2 g/m^2 (dry). Tensile strength increased by 33%, ring crush by 48% and Scott bond (z-direction strength) by 7.5%. This is not a common approach for synthetic resins however.

7.6 Troubleshooting

7.6.1 Loss of Strength on Addition of Resin

No or negative increase in sheet strength is not uncommon when dry strength resins are used for the first time. Due to the charged nature of the polymer, increases in fines and filler retention can often be measured. This increased sheet loading will have an adverse effect on strength that is only partly offset by the addition of the resin. Increasing the dosage further should overcome this, so long as the charge balance is not disrupted (see above). Increased fibre flocculation can have a similar effect. Loss in formation may accompany this, although this effect may also be due to increases in drainage.

On tissue machines, the addition of a polyacrylamide will frequently increase the adhesion of the sheet to the Yankee dryer. This will in turn increase the energy input required to crêpe the sheet off again, which in turn will increase the number of fibre bonds being broken. This is often desired by the tissue maker, since softness generally increases, and indeed may be the original purpose for using the resin. The adhesion appears at much lower dosages than are required for dry strength formation however, so a strength loss can be initially seen as the dosage is increased. Higher dosages should demonstrate strength improvements.

Interference by other additives will also prevent a dry strength resin from performing well. This is usually accompanied by changes in performance of the interfering agent, and may be identified from this. The usual cause of an effect such as this is addition of the dry strength resin too close to another additive, particularly of an opposing ionic charge. Fillers should be included in an investigation into such an effect, as they generally contain anionic dispersants in the slurry.

7.6.2 Poorer Sheet Formation

Increased flocculation, as mentioned above, is often the cause of this problem. This can generally be overcome by looking at the addition point of the resin, and ensuring that thorough mixing takes place after addition. This will break up the flocs and ensure acceptable formation.

Some dry strength resins offer a collateral benefit of increased drainage. This can also occur when the charge balance in a system changes. Increases in drainage can have an adverse effect on sheet formation due to the rapid consolidation of the fibre stock on the wire. This may be overcome by reducing the consistency of the flow box stock.

7.6.3 Loss of Charge Balance

Most (but not all) papermaking systems run under slightly anionic conditions. The addition of extra cationic materials may disrupt the balance in the wet-end, causing severe problems. These are generally evidenced through reduced drainage, poor dry strength development, lower retentions, increased foam levels, deteriorating formation, colour shifts in dyed grades, and loss of effectiveness of other functional chemicals. This may be corrected through adjustment to the addition levels of the chemicals, through a change from a cationic dry strength resin to an anionic product, or by adjusting the additions of a dual component system.

Two main techniques are used to measure the charge in a system. Electrophoresis techniques have been in regular use since the 1960s, although the sometimes difficult and monotonous procedures meant that the test did not gain widespread acceptance. Despite this, considerable research was carried out into the benefits of this approach on papermaking systems [14]. Nowadays streaming current potential has become the standard approach, due to its ease of use and reproducible results. It should however be pointed out that the two techniques do measure different parameters, and there is still a place for both technologies in a mill.

McKague and colleagues [15] put this approach to good use at Eastman Kodak. Their photographic paper contained a considerable number of cationic additives, and as a result, their system turned cationic. Although they were producing good quality paper under these conditions, they felt that by using their anionic dry strength resin to control charge on the anionic side, they could gain benefits.

The optimisation process took over a year, but resulted in considerable benefits. They increased production, achieving improvements of 25% on some grades, and also saw chemical additive savings of 14%.

7.7 Summary

Relatively low molecular weight polymers such as polyacrylamides or polyvinylamines are widely used as synthetic dry strength additives. These offer both economic and quality benefits to the papermaker that could not be obtained through increases in refining or wet pressing. Additions of active resin as low as 0.2% can achieve increases in dry strength of up to 50%. Collateral benefits can include increases in retention, drainage, machine runnability, productivity and reduced raw material costs.

References

[1] Linke, W.F. (1962), *Tappi*, 45, 326–333.
[2] Mortimer, D.A. (1991), *Polymer International*, 25, 29–41.
[3] Reynolds, W.F., Wilson, L.R., Thomas, W.M., Woodberry, N.T., Barthel, J.C. and Landes, C.G. (1957), *Tappi*, 40, 839–846.
[4] Reynolds, W.F. (1961), Some recent research on accostrength resin 2386, *Tappi*, 44(2).
[5] Cytec Industries product data sheets.
[6] Finneman, M.N. (1952), *Tappi*, 35(7), 320.
[7] Farley, C.E. (1988), Polyacrylamide Dry Strength Resins Applications, Tappi wet and dry strength short course.
[8] Cooper, R. (1990), Styrene-acrylate polymer additives for strength enhancement of paper. In *Speciality Chemicals for the Paper Industry*, PIRA Conference Proceedings.
[9] Neal, C.N. (1988), A review of the chemistry of wet strength development, *Tappi Wet and Dry Strength Short Course Proceedings*.
[10] Farley, C.E. (1988), Glyoxalated polyacrylamide wet strength resin, Tappi Wet and Dry Strength Short Course Proceedings.
[11] Farley, C.E. (1994), Glyoxalated polyacrylamide wet strength resin, Wet-Strength Resins and Their Applications, *Tappi*.
[12] Linke, W.F. (1962), *Tappi*, 45, 326–333.
[13] Strazdins, E. (1984), Chemical aids can offset strength loss in secondary fiber furnish use. *Pulp and Paper*, March.
[14] Strazdins, E. (1977), *Tappi*, 60(7), 113–114.
[15] McKague, J.F., Etter, D.O., Pilgrim, J.O., Griggs, W.H. (1974), *Tappi*, 57(12).
[16] Pelton, R., Hong, J. (2002), *Tappi*, 1(10).
[17] Esser, A., Rubenackker, M., Schall, N., Haffke, R. (2005), *Ipw* (10).
[18] Jenkins, S. (1995), The improvement of dry strength by synthetic polymers, in Au, C.O., Thorn, I. (eds.) *Applications of Wet-End Paper Chemistry*, Blackie Academic and Professional.

Chapter 8
Wet-Strengthening of Paper in Neutral pH Papermaking Conditions

M.T. Crisp and Richard J. Riehle

8.1 Introduction

The ability to wet-strengthen paper is an important requirement of the paper industry because wet strength agents allow for the manufacture of a variety of paper grades and products. The most significant area of application is in the manufacture of hygiene grades of paper, including hand towels and facial tissues, kitchen towels and serviettes, and other tableware products. Another important area of application is in packaging grades, including liquid packaging board for fruit juice and milk cartons; carrier bags and paper sacks; and corrugated cardboard boxes for transportation of fruit, meat and other items. Speciality grades, such as tea bags and coffee filters, bottle labels and currency paper, and wallpaper and poster paper also use wet strength agents. In fact, any grade of paper that needs to maintain a degree of strength and integrity after becoming wet involves the application of a wet strength agent.

8.1.1 Definition of Wet Strength

Wet strength is a measure of paper's ability to maintain its dry strength when it is wet. Paper has wet strength when it retains at least 10% of its original dry strength after it has been fully soaked in water. Paper can become fully soaked after just a few seconds of immersion in water, as is the case of tissue and towel grades, or it can require several hours of immersion, as is the case for some grades of liquid

M.T. Crisp
Ashland Hercules Water Technologies, Noordweg 9, 3336, LH Zwijndrecht, The Netherlands
e-mail: mcrisp@ashland.com

R.J. Riehle
Ashland Hercules Water Technologies Inc., Research Center, 500 Hercules Road, Wilmington, DE 19808, USA
e-mail: rjriehle@ashland.com

I. Thorn and C.O. Au (eds.), *Applications of Wet-End Paper Chemistry*,
DOI 10.1007/978-1-4020-6038-0_8, © Springer Science+Business Media B.V. 2009

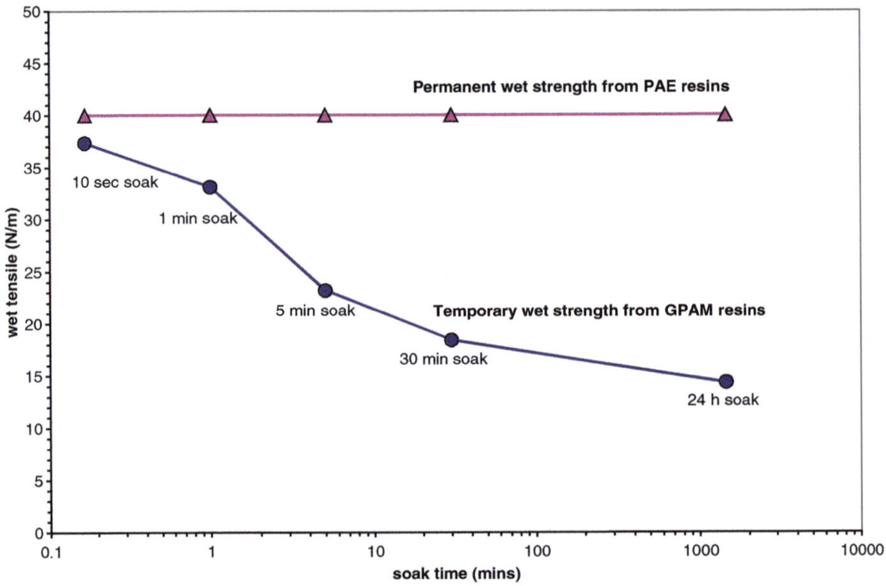

Fig. 8.1 Permanent versus temporary wet strength

package board, especially when other properties of the paper need to be overcome first, such as sizing. Typically, in laboratory testing, paper is soaked for 2 h in ambient temperature conditions before testing for strength properties.

Where the strength property is constant and independent of the length of time of soaking, the paper is considered to have 'permanent' wet strength. Where the initial wet strength, after the paper is fully soaked, is high but is then found to decrease with prolonged soaking, the paper is considered to have 'temporary' wet strength (Fig. 8.1). Temporary wet strength (TWS) can be an important property of the paper and is useful where easy reworking of broke, without the use of re-pulping aids, is required in a paper mill. TWS is also useful for specific grades of paper such as toilet tissue, where a high initial wet strength property is required during use, but the paper must be broken down easily after it is discarded.

8.1.2 History of Wet Strength Resins

The earliest known processes for imparting wet strength involved heating the paper to high temperatures or drying the paper after soaking it in dilute sulphuric acid. Later it was demonstrated that wet strength properties could be achieved by impregnating paper with formaldehyde. In the late 1930s and early 1940s, water-soluble synthetic polymers became commercially available that could provide wet strength properties. Since then the use of these 'wet strength resins' has grown rapidly and with it the manufacture and use of different wet-strengthened grades of paper.

The first commercial wet strength resins were based upon urea-formaldehyde (U-F) chemistry but were modified by the introduction of anionic and cationic groups into the polymer backbone to render them water-soluble and substantive to paper fibres. The underlying chemistry of these polymers meant that they could only function, practically, as wet strength agents in acidic papermaking conditions. The development of melamine-formaldehyde (M-F) resins supplemented the early technology. This M-F class of resins gained commercial acceptance because of its greater flexibility and versatility. M-F resins operate at a higher pH range than U-F resins but still in acidic papermaking conditions.

The change from acidic to neutral/alkaline pH papermaking conditions led to the development of polyaminopolyamide-epichlorohydrin (PAE) polymers in the late 1950s. In a neutral pH papermaking environment, these resins imparted higher levels of wet strength than U-F and M-F polymers. Furthermore, given the benefits of operating a paper machine in this pH range, such as greater natural fibre strength, reduced machine corrosion and better slime control, PAE resins became an overnight commercial success.

Glyoxalated polyacrylamide (GPAM) resins were developed and introduced in the 1960s. These were able to function in a near neutral pH environment. However, the level of wet strength imparted by GPAM technology was less than PAE resins at the same dosage level and the wet strength was not always permanent due to one of the chemical mechanisms involved. GPAM technology has had more success providing dry strength than wet strength to paper. More recently, the non-permanence of the GPAM resins' wet strength was exploited by modifications to the structural and functional characteristics. A range of GPAM resins is now being marketed and used as temporary wet strength (TWS) agents.

Today the wet strengthening of paper is done almost entirely in neutral/alkaline pH papermaking conditions. However, the wet strengthening of paper in acidic pH papermaking conditions is still necessary in some grades that often require additional functional chemicals, such as rosin sizing emulsions for sack paper, to provide sizing and frictional properties. For such applications and conditions, U-F and M-F resins are the only viable resin choices to provide wet strength. However, concerns remain about the presence of free formaldehyde in these resins, so wherever possible PAE resins are used. This shift from formaldehyde-containing resins to PAE resins continues not only in papermaking but also in other wood-related industries. Due to the inherent benefits of papermaking in a neutral pH environment, PAE resins are very much the wet strength resin of choice.

8.2 Polyaminopolyamide-Epichlorohydrin (PAE) Wet Strength Resins

8.2.1 Basic Chemistry and Classification of PAE Wet Strength Resins

Polyaminopolyamide-epichlorohydrin (PAE) wet strength resins are prepared by a two-step process that involves the following (Fig. 8.2):

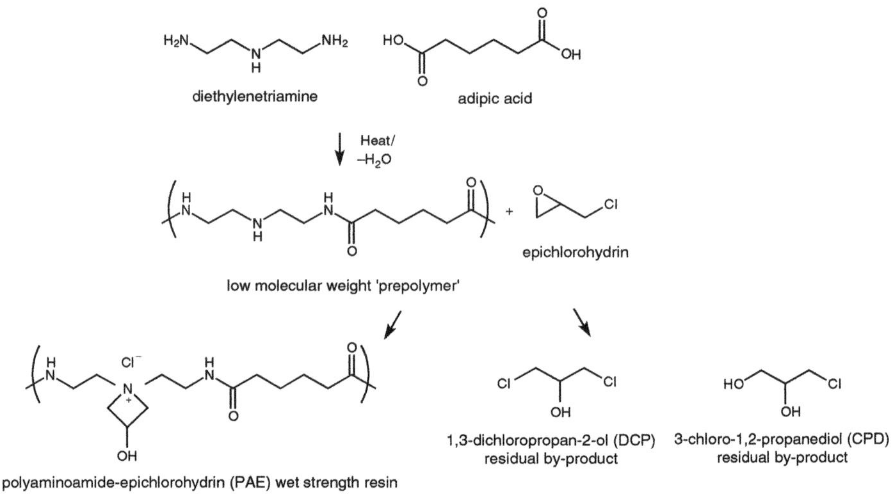

Fig. 8.2 Manufacture of PAE wet strength resins

1. The condensation reaction between a polyalkylenepolyamine (usually diethylenetriamine) and a dibasic acid (usually adipic acid) to form a lower molecular weight polyamide that contains a number of secondary amine sites within the backbone
2. The treatment of this lower molecular weight polyamide with epichlorohydrin, principally by reaction at these secondary amine sites, to form the key cationic, reactive functionality 3-hydroxyazetidinium chloride and to develop further the molecular weight of the PAE resin

Using epichlorohydrin to develop the structural and functional properties of the resin also results in the production of residual by-products dichloropropanol (DCP) and chloropropandiol (CPD). Over the years, these residual by-products have come under scrutiny from various regulatory bodies. The efforts by chemical manufacturers to reduce the levels of DCP and CPD can be used to classify the different grades of PAE resins available.

Products containing more than 1,000 ppm of DCP are now referred to as Generation 1 PAE wet strength resins. The original development of this technology, in the late 1950s, would fall into this class.

Products containing less than 1,000 ppm of DCP are now referred to as Generation 2 PAE wet strength resins. Development of this class of resins was started in the late 1980s and involved a combination of formulation changes and significant process chemistry that were designed to use more efficiently the epichlorohydrin that was added to the process. One consequence of these changes in the early development of this class of resins was a reduction in the relative amount of azetidinium functionality, compared to Generation 1 resins, and the subsequent negative impact this had on the performance of the Generation 2 resins in papermaking applications. To this day, development work on the Generation 2 wet strength resins continues in an effort to attain the higher levels of azetidinium functionality.

Products containing less than 10 ppm of DCP are now referred to as Generation 3 PAE wet strength resins. This class of resins meets the strict requirements of residual DCP and CPD levels permitted in some final paper products, such as liquid packaging board (LPB) and tea bag paper. Development of this class of resins was started in the 1990s, and usually involves the application of a post-manufacturing 'cleaning' technique that removes residual DCP and CPD from the product.

8.2.2 General Mechanisms of Wet Strength Development

Irrespective of its generation, the ability of a PAE resin to impart wet strength to paper is dependent upon two properties:

- The molecular weight of the resin and
- The amount of azetidinium functionality

A reduction in either of these properties will mean that the resin will not provide the maximum level of wet strength for the dosage level being used. The age of the resin can determine the amount of azetidinium functionality present in and the molecular weight of the resin at the time of use. Azetidinium functionality decreases as the resin ages due to hydrolysis mechanisms. Hydrolysis of the linkages in the backbone of the resin due to highly acidic stabilization conditions can result in loss of molecular weight. Chemical manufacturers provide recommendations for the shelf life of a product, beyond which the resin should not be used if wet strength targets are to be attained. Therefore, the level of reactive functionality and the structure of the polymer, at the time of use, must be considered when applying a PAE resin. Furthermore, factors specific to the papermaking system that can affect these properties must be considered when applying a wet strength resin.

A PAE resin must be retained on the surface of the paper fibre before it can impart wet strength to the paper. The cationic nature of PAE resins makes them naturally substantive to the anionic sites of paper fibres. Any factors that interfere with this initial mechanism will result in poor retention of the PAE resin and, consequently, lower than expected levels of wet strength in the paper.

Once on the paper fibre, the PAE resin must do at least one of four things:

- Add to or strengthen existing bonds
- Protect existing bonds from water
- Form bonds that are insensitive to water
- Form a network of material that physically entangles these bonds

When using PAE resins, two mechanisms explain the development of wet strength in paper (Fig. 8.3). The first is a 'preservation' or 'restriction' mechanism but is more often referred to as 'homo-crosslinking'. After adsorption onto the fibre surface, the azetidnium functionality of resin reacts with amine sites elsewhere on the backbone of the resin, effectively forming a self-crosslinked network that physically

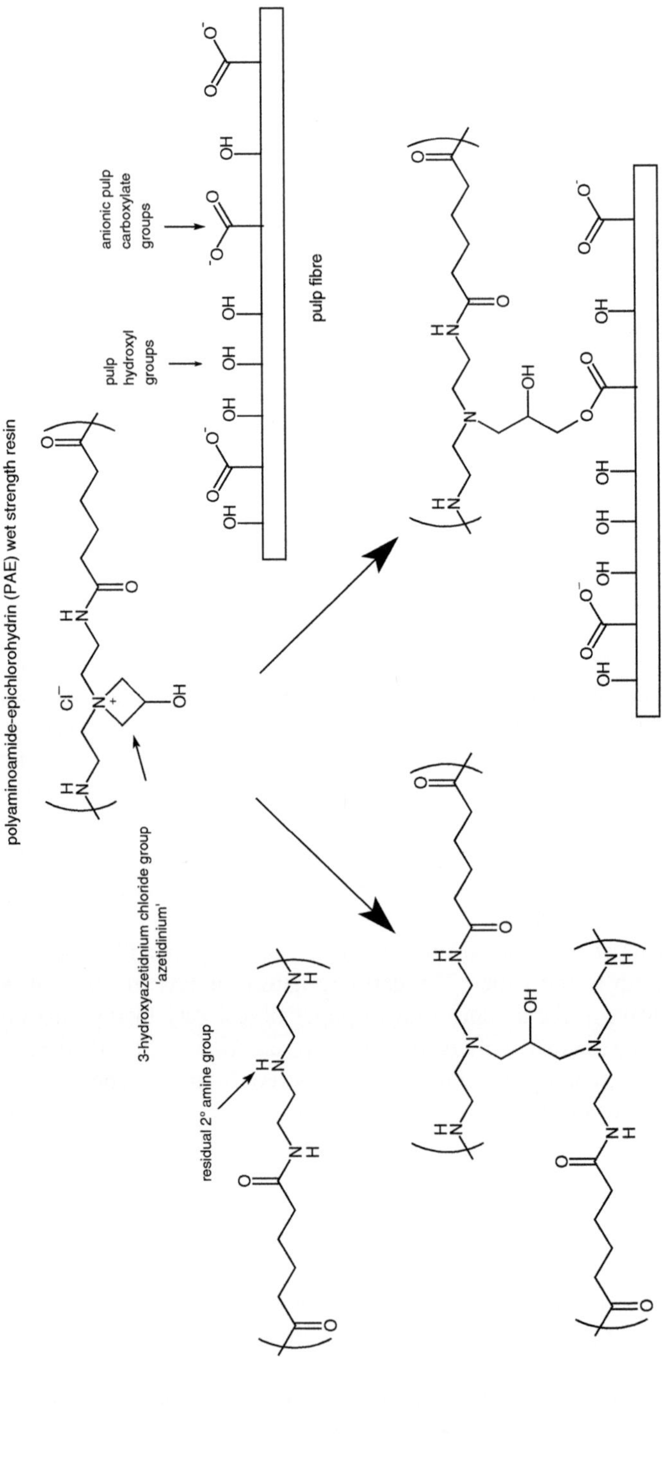

Fig. 8.3 Wet strength mechanisms of PAE resins

entangles fibres as the paper dries. The rate of the crosslinking reactions which form the network is greatly influenced by:

- System pH
- Local concentration of resin at the fibre as water is removed
- System temperature

When the paper is re-wetted, this resin network restricts the re-hydration and swelling of the fibres. The network also acts to protect the hydrogen bonds in the fibre-to-fibre contact area.

The second mechanism is a 'reinforcement' or 'new bond' mechanism but is more generally known as 'co-crosslinking'. Here the azetidinium functionality reacts with functional groups on the fibre surface, typically carboxylate functionality, to form covalent bonds. The resin is polymeric; therefore, it is capable of using this property to form chemical bonds with the surfaces of several fibres, thereby effectively crosslinking fibres together. When the paper is re-wetted, these chemical bonds will remain even after the water has disrupted all the naturally occurring hydrogen bonds. This mechanism tends to lead to a higher level of wet strength.

8.2.3 Application Guidelines for PAE Wet Strength Resins

8.2.3.1 Handling of PAE Resins

The development of cleaner Generation 2 and Generation 3 technologies and the development of higher solids containing PAE resin products have necessitated the stabilisation of PAE resins at low pH conditions. Typically, pH values less than 3.0 are required at the time of manufacture. The use of epichlorohdyrin to generate cationic azetidinium functionality also results in the formation of chloride counter ion. In the low pH environments of today's PAE resins, 'chloride ion induced corrosion' can present a major problem. The problem can be particularly acute at the welds between metal pipes. Therefore, the fabric of construction used for storage tanks and piping requires special attention. Corrosion-resistant plastics and metal alloys, such as CPVC or 316L, should be considered for addition points and unloading lines, whilst fibreglass reinforced plastic should be used to line product storage tanks.

To control dosing of the PAE resin to the papermaking system, metering pumps and mini magnetic flow or mass flow meters fabricated from 316L stainless steel and fitted with Teflon or Viton-wetted seals are most preferred. Progressive cavity or centrifugal pumps are preferred because they are low shear and impart less heat to the PAE resin.

8.2.3.2 Factors Affecting the Retention of PAE Resins

To provide wet strength, a PAE resin must be retained on the surface of the paper fibre. The principle mechanism of this retention is the ionic attraction between the anionic carboxylate groups on the fibre surface and the cationic azetidinium groups of the resin. The anionic charges are effectively neutralised as the resin adsorbs onto the fibre surface. This reduces the effective surface charge area. As more resin adsorbs, the effective surface charge declines to a point where the surface is effectively saturated; the fibre cannot retain any additional PAE resin. In practical terms, this leads to a maximum level of resin that can be retained and capable of imparting wet strength to the final paper product. If the dosage level of PAE resin exceeds this point then higher levels of wet strength will not be attained and the excess PAE resin will pass into the white water of the papermaking system and cause problems.

Increasing the carboxylate content of the fibre increases the retention of PAE resins (Fig. 8.4). The level of carboxlyate content is very much a function of the pulping and bleaching processes used in the manufacture of the fibre. Generally the pulps that contain the highest carboxylate levels and are the most responsive to PAE resins are unbleached Kraft pulps (UBK) followed by bleached hardwood Kraft pulps, bleached softwood pulps, recycled secondary fibre and then sulphite pulps. Elemental chlorine free (ECF) bleaching of the pulp tends to result in a higher level of residual carboxlyate groups on the fibre than total chlorine free (TCF) bleaching.

Refining the pulp increases the surface area of the fibre, exposing more carboxylate sites capable of attracting the PAE resin. This allows a greater level of PAE resin to be adsorbed onto the fibre before reaching the maximum level of PAE resin. The effect is most pronounced in very responsive pulps such as UBK. However, refining also increases the levels of fines in a paper furnish. PAE resins absorb equally well onto

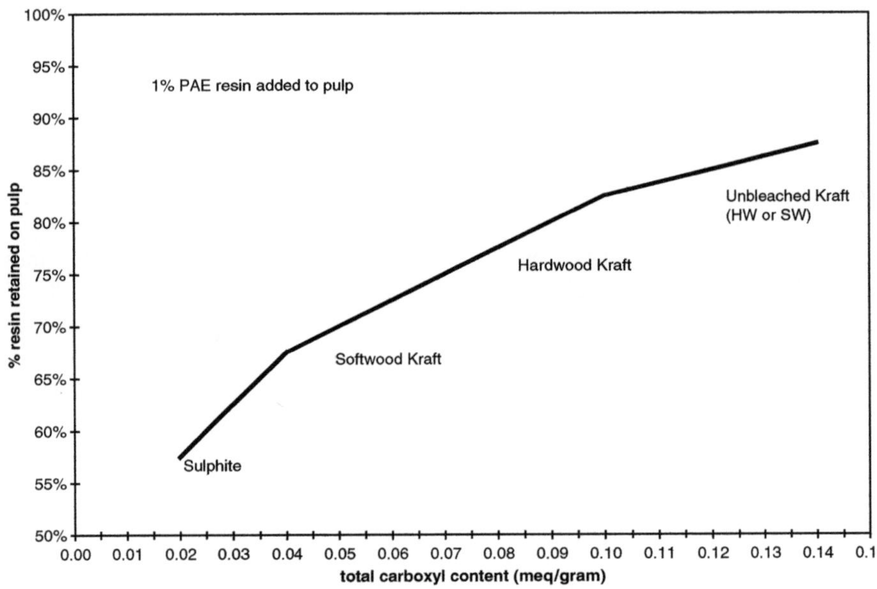

Fig. 8.4 Effect of pulp carboxyl content on PAE resin retention

the anionic sites of fines as well as fibres. If the sheet does not retain the fines, a significant portion of the PAE resin will be lost, resulting in poor wet strength levels. Furthermore, refining can influence other parameters, such as drainage, that are critical to the smooth, efficient running of a paper machine.

In general, adding the resin to the longest fibre or most reactive pulp attains the best wet strength levels. The resin should be distributed evenly across the pulp to maximise its presence in the 'fibre-to-fibre' contact areas. Diluting the product to about 1% (through a static mixer on its way from the storage tanks to the point of addition) in the papermaking system, results in good distribution of the resin. Dilution of the resin requires fresh water, free of any residual chemicals and components from the papermaking system that might interact in a detrimental fashion with the PAE resin prior to contact with the fibre.

To aid distribution further, the resin is usually added to the thick stock. Good mixing is important to aid distribution and so the resin is usually added at the machine chest pump to help with this. Some mills use a static mixer in the stock flow after the point of resin addition. In some cases, splitting the incoming resin flow and using an annular ring with multiple dosing points aids resin distribution. Adding to the thick stock also allows the resin to be added early in the papermaking system, thereby providing sufficient time for the resin-retention mechanism to occur. However, if the stock temperature is too high and/or the contact time is too long, then hydrolysis of the resin can occur. Hydrolysis affects both functional and structural properties of the resin and therefore the level of wet strength that it imparts. To maintain the PAE resin in its optimal condition, the stock temperature and the contact time are considered.

Aside from low wet strength values, the consequences of poor retention of the PAE resin onto the fibre or of fines that have absorbed the PAE resins are:

- Foam – The chemistry and cationic nature of PAE resins provide a very good surfactant property for the formation and stability of foam. Overuse of PAE resins to attain a desired strength target often exacerbates this problem and results in excess resin in the white-water system. The presence of foam is often a sign of poor PAE resin retention.
- Felt filling – Excess PAE resins, either on poorly retained fines or in the water phase, can lead to felt filling, which is a major concern of paper manufacturers. Effective resin retention strategies and good thorough mechanical cleaning of the felts manage the problem. However, ultimately, the machine must be shut down to implement a chemical cleaning programme.
- Adverse effects on Yankee coatings – During the manufacture of tissue and towel grades, contact with a large Yankee cylinder dries the paper web. The paper is 'crêped' off the Yankee cylinder using a doctor blade. The action of crêping disrupts the fibre-to-fibre bonding within the paper and can fracture the fibre itself, which provides the soft hand-feel quality required for some grades and imparts additional bulk and absorbency properties to other grades. Unfortunately, the crêping action also results in wear of the doctor blade and of the surface of the Yankee cylinder. To avoid the need to regrind or replace a Yankee cylinder,

the surface is protected by the application of a multi-chemical 'coating package'. This coating package balances the adhesive properties required to pick up and adhere a wet sheet to the surface of a Yankee cylinder and the release properties required at the doctor blade to allow the crêping action to occur. The coating package provides a soft coating that allows the doctor blade to pass easily through for a smooth crêping action and, yet adequately protects the Yankee cylinder surface. The adhesive components are provided by water-soluble polymers similar in chemical nature of PAE resins, but unable to provide wet strength in their own right. These polymers are capable of reacting with PAE resins. Consequently, a poorly retained PAE resin that is not fixed within the matrix of the wet web can migrate to the coating and crosslink with the adhesive polymers. This hardens the coating and impairs the crêping action.

8.2.3.3 Factors Affecting the Reaction of PAE Resins

After the PAE resins are retained on the fibre and the fibres are consolidated into a wet web in the wire section of the paper machine, the resins impart wet strength to the paper via a homo-crosslinking and/or a co-crosslinking mechanism. The latter mechanism imparts a higher level of wet strength than the homo-crosslinking mechanism. Paper manufactured using pulps containing high levels of carboxylate groups experiences a greater proportion of the co-crosslinking mechanism.

Irrespective of which mechanism imparts the wet strength, both require the application of heat to dry the paper and to drive the chemical reactions. In most cases, the maximum level of wet strength obtained from a PAE resin does not fully develop on the paper machine. The paper requires ageing to allow the wet strength to develop fully. However, the dryer temperatures can influence the level of wet strength obtained immediately off the paper machine and ultimately the maximum level of final wet strength. The higher dryer temperatures result in lower sheet moisture levels, bringing fibres and resin closer together, thereby improving reactivity and reaction rates. At higher dryer temperatures the wet strength development of the paper is improved both 'off-machine' and with ageing.

8.2.3.4 Effect of Water Quality

The quality of the water in the wet-end of the papermaking system can have a large bearing on the performance of the PAE resin. Whilst dilution of the resin from the storage tanks to the chosen point of addition requires fresh water, recycled water is used for preparation of the thick stock and subsequent dilutions to the thin stock prior to formation of the wet web. The properties of the recycled water can determine the effectiveness of the PAE resin.

The pH value of the water can have one of the biggest effects on the resins' effectiveness. PAE resins work best when the value of the papermaking system pH is between 6.5 and 8.5. No beneficial effect on wet strength performance occurs when operating at higher pH values. However, lower pH values, which affect both the retention mechanism and the subsequent crosslinking reactions, have a detrimental impact on the ability of the resin to impart wet strength. In terms of retention, lower pH values can actually increase the overall cationic charge of a PAE resin because residual amines can be protonated to create a cationic charge. However, carboxylate groups on the fibre surface also protonate to their carboxylic acid form, thereby reducing the overall anionic charge of the fibre. This makes the fibre less substantive to the cationic resin, thereby affecting the original retention mechanism. In terms of reaction, protonation of the residual amine groups on the backbone of the resin due to a lower pH value inhibits the homo-crosslinking reaction and, therefore, the formation of the resin network in and around the fibres. Lower pH values also mean that carboxyl groups on the surface of the fibres are not dissociated to their carboxylate forms, rendering them incapable of forming covalent bonds with the azetidinium functionality of the resin.

The total alkalinity of the water, in the form of carbonate and bi-carbonate ions, has a beneficial impact on the performance of PAE resins. Increased alkalinity improves fibre swelling, which exposes more reactive carboxylate sites, improves overall resin retention onto the fibre and increases the rate of the co-crosslinking reaction. Furthermore, carbonate and bi-carbonate act as catalysts for the ring-opening reaction of the azetidinium groups increasing reactivity.

The hardness of the water can be problematic for PAE resins. Divalent cations such as calcium and magnesium can tie up pulp anionic (carboxyl) sites and inhibit retention of the cationic PAE wet strength resin. Hardness levels up to 150 ppm usually present no problems.

The presence of free chlorine in the water can break linkages in the backbone of a PAE resin, thereby decreasing its molecular weight and its ability to impart wet strength. Low molecular weight resin fragments cannot build a good crosslinked polymer structure in the paper. Typically, PAE resins can tolerate a maximum level of 1 ppm of free chlorine. Above this level, a decrease in wet strength occurs. The use of 'antichlors', such as sodium sulphite, bisulphite and thiosulphite, can control free chlorine levels. These reducing agents act to convert free chlorine (and/or hypochlorite ions) to inorganic chloride ions. However, excessive levels of these ions can also have a detrimental impact on PAE resins. Sulphite ions can react with azetidinium groups to form pendant sulphonate species on the resin backbone, thereby reducing the overall reactivity of the resin. PAE resins usually can tolerate up to 50 ppm of sulphite ion. Above this level, a decrease in wet strength is noticeable.

Soluble anionic salts cause high conductivity of water. Whilst these salts usually do not interfere with cationic resin retention, they are accompanied by soluble anionic organic species often referred to as 'anionic trash'. These species, which can be naturally occurring humic or pectic acids or pulping and bleaching residues, such as ligninsulphonates, readily form insoluble complexes with the cationic PAE resin. These complexes have little or no charge and severely impair the adsorption of the resin onto the fibre. This can result in a considerable loss of wet strength. Washing the pulp, where possible, removes a great deal of these anionic species.

8.2.3.5 Using Additional Polymers to Enhance PAE Resin Performance

As reviewed earlier, the ability of a PAE resin to impart wet strength to paper is dependent upon the following:

- Age and quality, specifically the structural and functional characteristics, of the resin at the time of use
- Wet-end papermaking conditions, which can determine the retention of the resin onto the fibre and the subsequent reactions of the resin that develop wet strength
- Drying conditions, which determine the rate of reaction of the PAE resin

Using additional polymers can overcome some of the factors that impede wet strength development. These polymers can even 'boost' the performance of a PAE resin in ideal conditions.

Where high levels of anionic trash are present, and it is not possible to clean the pulp by washing, fixing agents or 'anionic trash collectors' (ATC) may be used to control the overall conductivity and charge of the system. ATCs are typically polyamine-epichlorohydrin polymers that are added to the papermaking system before the PAE resin. The cationic charge of these ATCs is significantly higher than the cationic charge of PAE resins. Therefore, ATCs very effectively form neutral complexes with anionic species, which then clean the water of the papermaking system. Whilst ATCs are very useful at removing species that would impair PAE resin retention, they must be used carefully to avoid over-cationization of the system. Use of an online charge measurement tools helps with the metering of ATCs, with −20 and −50 μEq/ml being the target ionic demand.

Aside from adding highly cationic polymers that form complex anionic species that would otherwise impair the adsorption of PAE resins to fibre, anionic polymers can be added to further increase the retention of PAE resins, resulting in increased wet strength performance. Carboxymethylcellulose (CMC) and anionic polyacrylamides (APAM) are typical examples of anionic polymers added to papermaking systems. Using CMC or APAM polymers improves PAE resin retention by creating more anionic sites on a PAE resin-saturated fibre. The anionic polymer interacts with the PAE resin already retained on the fibre, which allows the polymer to be retained as well. The retention of these secondary polymers introduces additional anionic sites that can in turn attract, interact with, and retain, more cationic PAE resin. This process results in the formation of a large polymer complex or coacervate (Fig. 8.5). Due to the higher molecular weight of CMC and APAM polymers this large polymer complex is able to bridge wider fibre-to-fibre spans, thereby improving wet strength performance. The anionic polymers are usually added to the thick stock after the addition of the PAE resin; good mixing is essential. The ratio between the cationic and anionic polymers is critical. If this ratio is incorrect, the polymer complex can form deposits within the machine. Additionally, adding either the PAE resin or the anionic polymer to the thin stock also can result in deposition. The choice between CMC and APAM polymers usually depends on factors not directly related to wet strength performance. CMC is usually provided as a dry powder, which requires handling of the product to prepare a 1–2% solution; APAMs are provided as aqueous

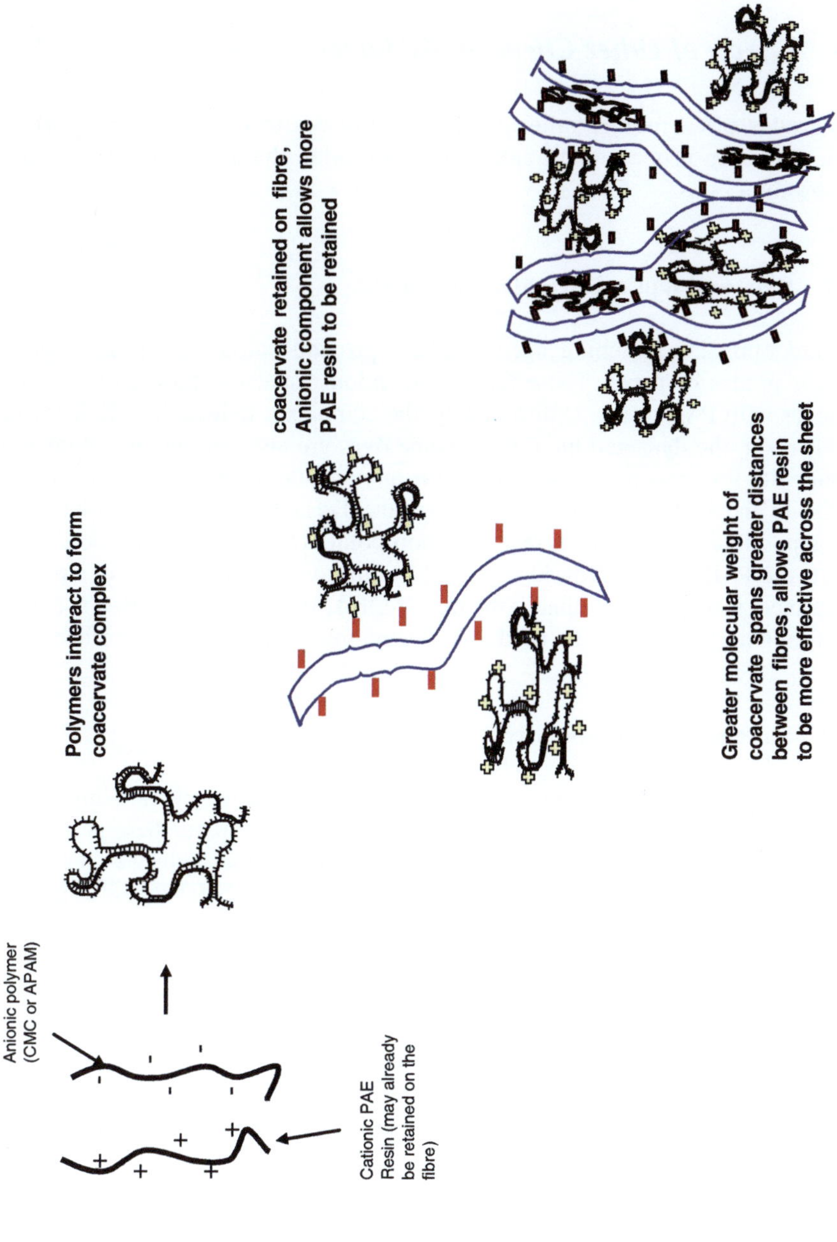

Fig. 8.5 Formation of coacervates

polymer solutions, which do not require handling or additional preparation. APAM polymers can produce a stiffer sheet, which is not desirable in the production of hygiene grades of paper. For those paper grades, CMC polymers are a better choice.

8.2.4 *Effects of Other Chemical Additives*

Paper grades often require other chemical additives to improve some part of the papermaking process or to enhance a particular characteristic of the final paper product.

8.2.4.1 Dyes and Optical Brightening Agents

Dyes and optical brightening agents (OBA) provide colour and brightness to particular grades of paper. These tend to be anionic species capable of forming complexes with PAE resins. Unfortunately, the complexes reduce the efficiency of the PAE resin, the dyes and the OBAs. Some dyes are also cationic in nature and compete with PAE resins for retention on the anionic sites of the fibre. When using dyes, OBAs and a PAE wet strength resin, keep the dye and OBA addition point as far away as possible from the PAE resin addition point. The order of addition should be, first, the PAE resin; followed by CMC to improve retention and wet strength performance, if needed; and finally the dye or OBA. As the dye, or OBA, is added last, a fixing agent may also be used to aid its retention onto the fibre surface.

8.2.4.2 Softeners and De-bonders

For hygiene grades of paper, bulk and surface softness characteristics are important. The action of crêping during manufacture of these grades imparts these properties to the paper. Adding softening and de-bonding agents further enhances bulk and softness. De-bonding agents are typically hydrophobic molecules that also have a cationic charge. They work by disrupting the fibre-to-fibre bonding sites, which increases bulk softness but reduces the overall strength of the paper. Since these agents are usually added to the wet-end of a papermaking system they compete with PAE molecules for retention at the anionic sites on the fibre. Careful consideration of the addition sequence is the best way to maximise the performance of the two additives. Softening agents are similar in chemistry to de-bonding agents, but softening agents act by lubricating the surface of fibres. This results in a decreased coefficient of friction of the paper and an increased perception of surface softness. Softening agents can be added to the wet-end of the papermaking system or they are applied to the surface of the paper with a suitable spray system. When added to the wet-end, softening agents interfere with fibre-to-fibre bonding, thereby impeding the development of wet strength by the PAE resin. When softening agents are sprayed onto the surface of either the wet web, if the web is suitably supported, or the dry sheet, during rewinding or converting operations, they interfere only minimally with the development of wet strength by the PAE resin.

8.2.4.3 Absorbency Aids

PAE resins impart wet strength to paper, but they inhibit absorbency. Using absorbency aids offsets the negative affect of the PAE resins. Absorbency aids are surface-active re-wetting agents and are added to the wet-end of the papermaking system. They usually require quite a long dwell time to ensure good contact with the fibre and to be effective. However, this also means they can interfere with fibre-to-fibre bonding and impede wet strength development by the PAE resin.

8.2.4.4 Sizing Agents

In some paper grades that require dimensional stability, such as liquid packaging board, a PAE wet strength resin is often used with a sizing agent. The specific requirements of the manufactured grade and the pH value of the papermaking system determine the appropriate sizing agent. The addition sequence of these very different functional chemicals is the key to ensuring maximum performance. In neutral/alkaline pH systems where reactive sizing agents such as alkyl ketene dimer (AKD) or alkenyl succinic anhydride (ASA) are used, the PAE resin is usually added first. As well as providing wet strength, the PAE resin acts as a 'promoter' resin for the sizing agent. In acidic pH systems, rosin-sizing agents, together with additional alum, are more likely to be used. In these instances, the sizing agent is added first, followed by the alum and then the PAE resin. Adding the PAE resin after the alum can be advantageous in 'dirty' papermaking systems because the highly cationic Al^{3+} ion acts as a scavenger for anionic trash, thereby cleaning the system, which aids retention of the PAE resin.

8.2.5 Choice of PAE Wet Strength Resins

Since the mid-1980s, PAE resin technology has been the subject of significant research and development investigation and investment by chemical suppliers. Regulatory requirements and environmental pressure advanced PAE resin technology from the early Generation 1 resins to the cleaner and widely used Generation 2 resins and, finally, to Generation 3 resins. Generation 1 resins contain very high levels (typically >1,000 ppm) of residual DCP. Generation 2 resins contain much lower levels (<1,000 ppm) of residual DCP. The need to meet specific recommendations pertaining to the levels of DCP and CPD that may be permitted in some paper products used in Europe has resulted in the development and use of Generation 3 resins. These resins typically contain less than 10 ppm of DCP.

All generations of PAE resins behave the same way when imparting wet strength; thus the same factors and variables presented earlier affect their performance. The choice of which class of PAE resin to use is determined by the cost-effective ability of the resin to impart wet strength, the environmental impact of the wet strength agent on the AOX levels found in the effluent of the mill, and the level of residual DCP and CPD permitted in the specific paper product. In some instances, the organochlorine (OX) content of the paper itself may also be a deciding factor in the choice of which PAE resin to use. For paper to be considered TCF the OX content must not exceed a value of 30 mg/kg.

In 2001, the German Federal Institute for Consumer Health Protection and Veterinary Medicine (BgVV) finalised their recommendations for the levels of 1,3-DCP and 3-MCPD allowed in the water extracts of finished paper products. This recommendation reads as follows:

> *1,3-Dichloro-2-propanol must not be detectable in the aqueous extract from the finished product (detection limit 2 μg/l). The transfer of 3-monochloro-1,2-propanediol into the water extract of the finished products must be as low as technically achievable; a limit of 12 μg/l must not be exceeded in any case.*

For paper grades such as tea bags, coffee filters and liquid packaging board, these limits on the water extracts translate back to limits in the paper of 50 ppb for 1,3-DCP and 300 ppb for 3-MCPD. Only Generation 3 technology meets these very demanding requirements. However, a revision in the analytical methodologies used for tissue and towel grades means that although the same limits apply for 1,3-DCP and 3-MCPD in the aqueous extracts, the levels permitted in the actual paper product are an order of magnitude higher. The early Generation 2 resins are not capable of meeting these limits. The newer technologies, however, including the higher solids containing and more efficient Generation 2 resins, easily meet these limits.

The increase in the use of TCF grades of kitchen towel means that an additional class of technology is required. This class falls between Generation 2 and Generation 3 and is sometimes referred to as Generation 2.5. These resins still contain several hundred ppm of DCP and CPD, but they are capable of meeting the strictest requirements for DCP and CPD levels permitted in paper products. They also allow for low levels of OX in the paper to be attained, so that it can be classified as TCF quality.

8.3 Glyoxalated Polyacrylamide (GPAM) Wet Strength Resins

Whilst PAE resins predominate, glyoxalated polyacrylamide (GPAM) resins are the next most significant technology used to impart wet strength to paper. GPAM resins distinguishes themselves from PAE resins in that one of the mechanisms for wet strength development is the formation of chemical bonds with paper fibres that are reversible when the paper is rewetted. This attribute allows paper manufactured using GPAM wet strength resins to be reworked and recycled more easily. Recent developments in the technology have focused on manipulation of the resin structure and functionality levels to increase the initial short-term wet strength but then to promote rapid decay of the wet strength. This grade of GPAM resin is now used extensively to provide TWS to hygiene grades of paper, particularly toilet tissue paper. GPAM resins are also used in recycled liner board (RLB) to provide dry strength, to improve retention and drainage, and to increase machine speed.

8.3.1 *Basic Chemistry and Wet Strength Mechanisms*

The manufacture of GPAM and PAE resins is similar. First, a relatively low molecular weight co-polymer of acrylamide and diallyldimethylammoniumchloride is prepared. This 'prepolymer' is then reacted with glyoxal to form the desired GPAM resin (Fig. 8.6). The glyoxal performs two functions:

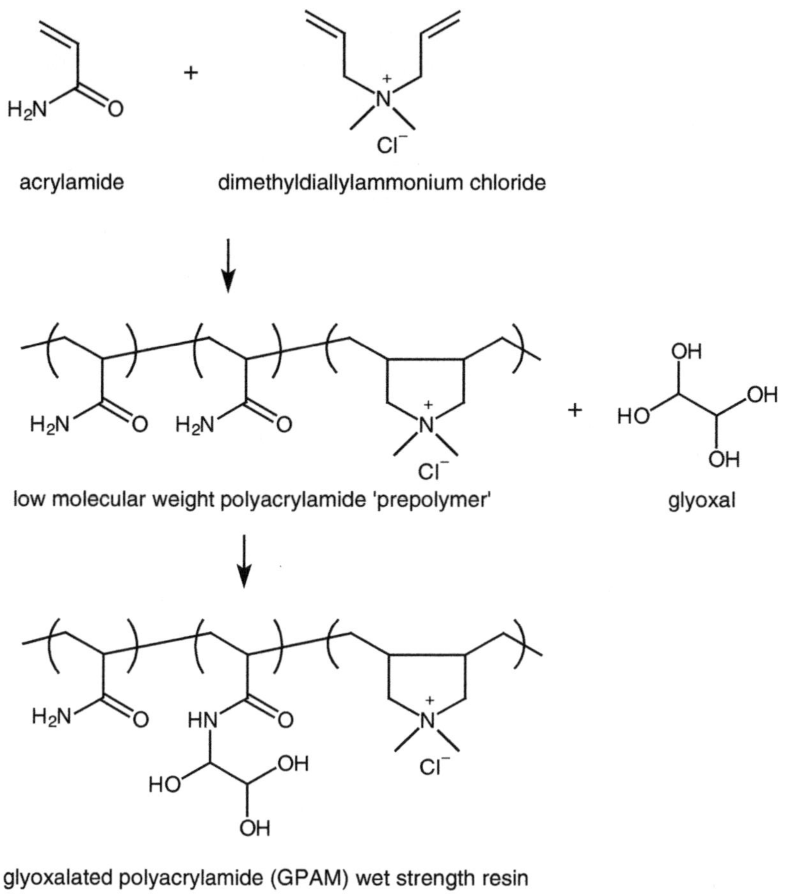

Fig. 8.6 Manufacture of GPAM wet strength resins

1. Reacting once to give the active polymeric aldehyde functionality
2. Crosslinking the prepolymer to build to the target molecular weight

The chemical mechanisms by which GPAM technology imparts wet strength to paper are also analogous to those described for PAE resins (Fig. 8.7). The major mechanism by which GPAM resins provide wet strength is the formation of hemi-acetal functionality between the aldehyde groups of the GPAM resin and the hydroxyl groups present on the pulp fibre. These bonds are formed as the wet paper web is de-watered and dried to paper product. This mechanism is analogous to the co-crosslinking reaction mechanism between azetidinium functionality and pulp carboxylate groups described for PAE resins. However, the hemi-acetal linkage formed between the GPAM resin and the paper fibre is sensitive to hydrolysis. As paper is re-wetted, these linkages break, thereby resulting in a loss of wet strength as a function of soaking time. Changes to the resin structure and level of aldehyde functionality can promote a high initial wet strength value and then a rapid decay of this wet strength. This forms the basis of temporary wet strength.

GPAM resins also have a significant permanent wet strength component. The predominant mechanism for this is considered to be a self-crosslinking mechanism, similar to the homo-crosslinking mechanism described for wet strength provided by PAE resins. For GPAM resins, the polymeric aldehyde groups react with polymeric amide groups to form hemi-amidal linkages. This intermolecular crosslinking also accounts for the relatively short shelf life of GPAM resins. GPAM resins' wet strength effectiveness typically increases as viscosity increases until the resins gel. The short shelf life of GPAM resins has limited their commercial applicability.

8.3.2 Relative Cure Rate of Wet Strength Functionality

Aldehyde functionality is more reactive than azetidinium functionality (Fig. 8.8). Therefore, GPAM resins impart greater initial wet strength than PAE resins. To impart wet strength using PAE resins requires heat energy to dry the paper and to drive azetidinium reactions. The level of wet strength the paper develops 'off machine' is a function of the heat energy applied in the dryer sections of the paper machine. Thereafter, paper still requires ageing, which allows the PAE resin to further cure and to impart more wet strength. GPAM resins impart a substantial level of wet strength using very little heat energy; the paper needs only to be de-watered and dried at ambient temperatures. Ageing of the paper allows additional wet strength to develop. The need to remove only water to develop wet strength suggests that the co-crosslinking mechanism of hemi-acetal bond formation is predominant here. This is useful for paper grades that require a substantial amount of off-machine wet strength. It can also improve the runnability of a paper machine by imparting some degree of wet strength to the wet web during the pre-dryer sections of the papermaking process.

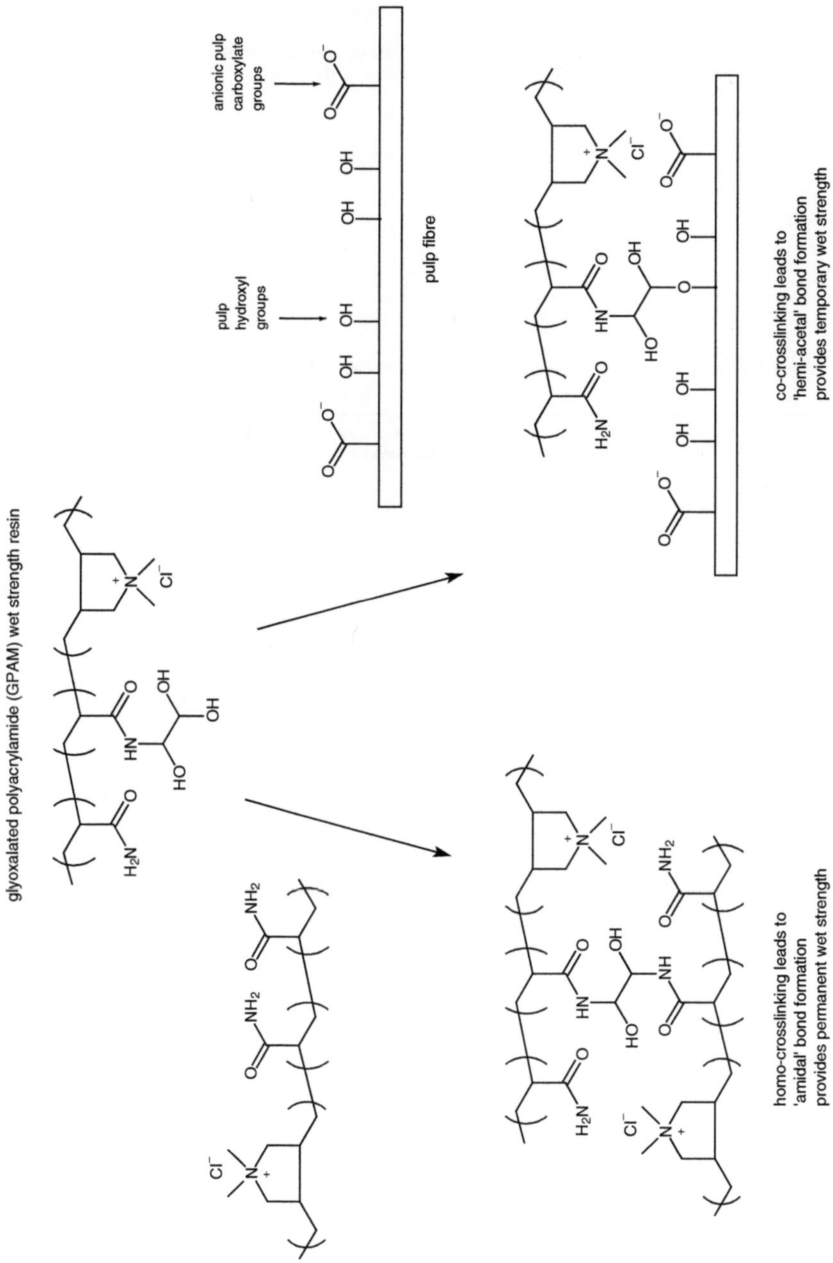

Fig. 8.7 Wet strength mechanisms of GPAM resins

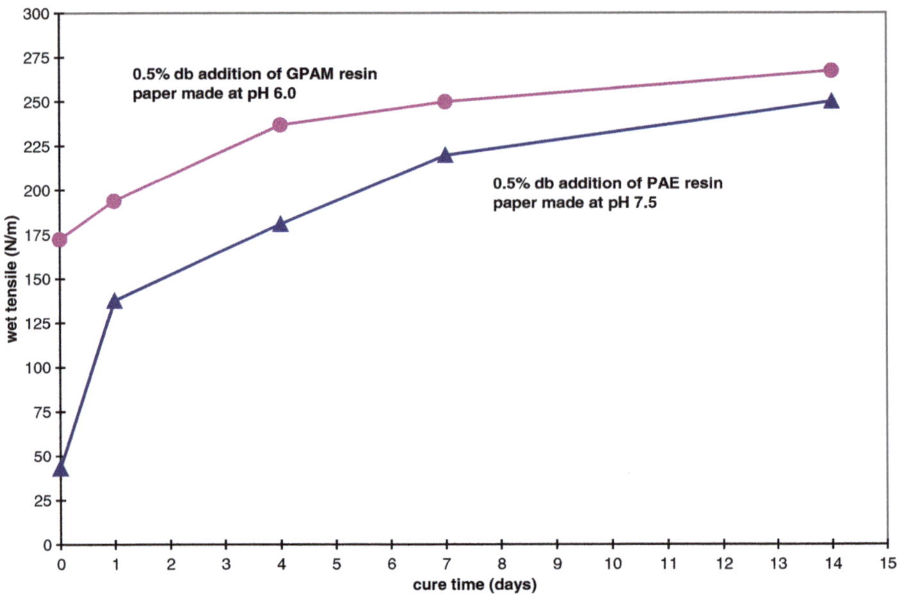

Fig. 8.8 Rate of cure of PAE resins versus GPAM resins

8.3.3 Application Guidelines for GPAM Wet Strength Resins

Many of the factors and practices described for the use of PAE resins apply to the use of GPAM resins. Typically, GPAM resins stabilise in acidic pH conditions, and because of the ionic chloride present in the composition, they should be handled and metered using the same guidelines recommended for PAE resins. The functional and structural characteristics of GPAM resins are not prone to decay with ageing, but GPAM resins are more prone to gel with prolonged storage. Dilution, which lowers solids, may extend the shelf life of the product. However, an increase in viscosity, which can also be an indication of increasing molecular weight, can prove detrimental to the decay characteristics of GPAM resins specifically designed to impart temporary wet strength. As cationic resins, they are retained on the paper fibres by similar mechanisms proposed for PAE resin retention. The retention mechanism is subject to the same factors that affect PAE resin retention and effectiveness, including point of addition to the stock, stock contact time and temperature, and interference from other anionic species. These also may be counter-acted using the same strategies employed for maximising PAE resin retention. GPAM resins are less cationic than PAE resins and, therefore, are more tolerant of some papermaking variables, such as increased water hardness and higher conductivity. Unlike PAE resins, the use of additional anionic polymers, such as CMC or APAM, does not significantly enhance GPAM resin performance. GPAM resins are especially sensitive to some papermaking variables.

8.3.3.1 Effect of pH

GPAM resins are capable of imparting wet strength to paper at near-neutral pH conditions but their wet strength capability decreases as pH values increase. This is especially critical when GPAM resins are used to provide temporary wet strength. The best rate of decay of the wet strength occurs when paper is manufactured at pH 6.0 (Fig. 8.9). When the pH value of the papermaking system is higher, more resin is required for the paper to attain the appropriate level of initial wet strength and the rate of decay is much slower. A likely explanation is that lower pH favours reaction with pulp hydroxyls and higher pH favours intermolecular crosslinking.

8.3.3.2 Effect of Alkalinity

Although alkalinity has a positive affect on wet strength imparted by PAE resins, it has a serious negative affect on GPAM resins' wet strength performance when alkalinity levels are above 100 ppm. The application of heat during drying of the paper causes the negative affect. If paper is air-dried, it develops a normal level of wet strength, even when the alkalinity is much higher than 100 ppm. Subsequent heating of this paper does not result in a loss of wet strength. Adjusting the papermaking system to slightly acidic pH values can offset or even eliminate the affect of high alkalinity.

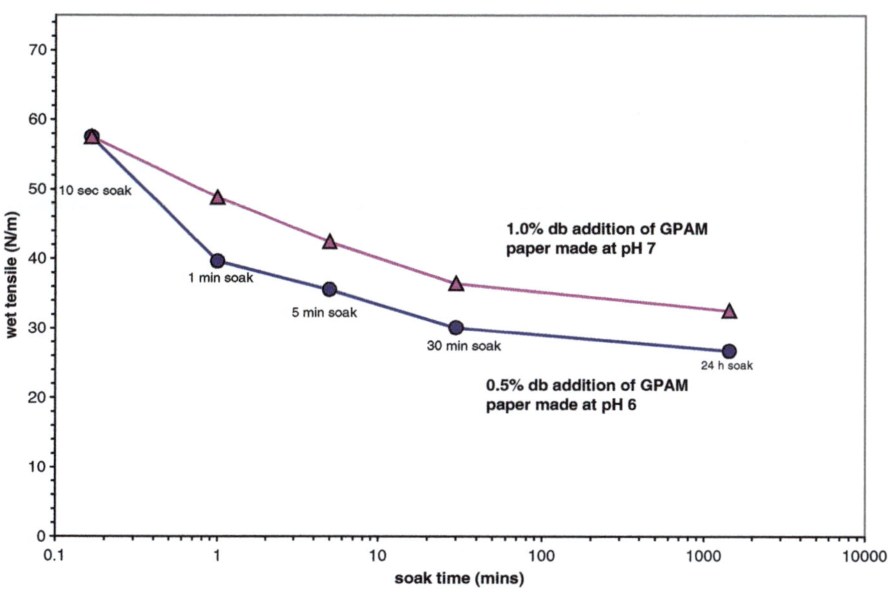

Fig. 8.9 Effect of pH on temporary wet strength performance of GPAM resins

8.3.3.3 Effect of Free Chlorine and Sulphite

Above a concentration of 1 ppm, free chlorine also has an adverse affect on the wet strengthening capability of GPAM resins. However, GPAM resins are especially sensitive to residual sulphite and bisulphite ions from 'antichlors' that may be used to control the level of residual chlorine. As little as 2 ppm of sulphite can have a negative impact on the wet strength capability of a GPAM resin. Bisulphite adducts are readily formed with the aldehyde groups in GPAM resins. This effectively 'blocks' the capability of the aldehyde groups to react with pulp fibres. Furthermore, the formation of these adducts creates anionic sites on the resin, thereby reducing the net cationic charge of the polymer. This lower cationic charge can adversely affect the initial retention of GPAM resins, thereby also having a negative effect on wet strength performance. Using modified GPAM resins that contain increased levels of cationic charge can offset the impact on retention.

8.3.3.4 Impact on Crêping

Whilst some GPAM resins are used as adhesives in crêping applications, generally they have been found to have a negative effect on the performance of established crêping adhesives. The aldehyde functionality of GPAM resins are capable of forming hemi-acetal and hemi-aminal linkages with the different adhesive components of a Yankee coating package, thereby causing the overall coating to become harder.

8.3.3.5 Impact on Absorbency

GPAM resins are very hydrophilic and have less of an impact on sheet absorbency than PAE resins. These resins can be used together to meet the overall wet strength requirements of a paper grade and to reduce the negative impact on absorbency that results from using only PAE resins. This avoids the need to use additional rewetting agents.

8.4 Polyvinylamine (PVAm) Wet Strength Resins

The wet strength capabilities of polyvinylamine (PVAm) resin were first reported in 1955. Subsequently, many researchers have sought a commercially viable manufacturing route to PVAm that allows for cost-effective use in papermaking applications. Today, PVAm is made by the hydrolysis of polyvinylformide; partially and fully hydrolysed polyvinylformamide is used increasingly in the manufacture of recycled liner board. Much of this growth has been at the expense of GPAM technology.

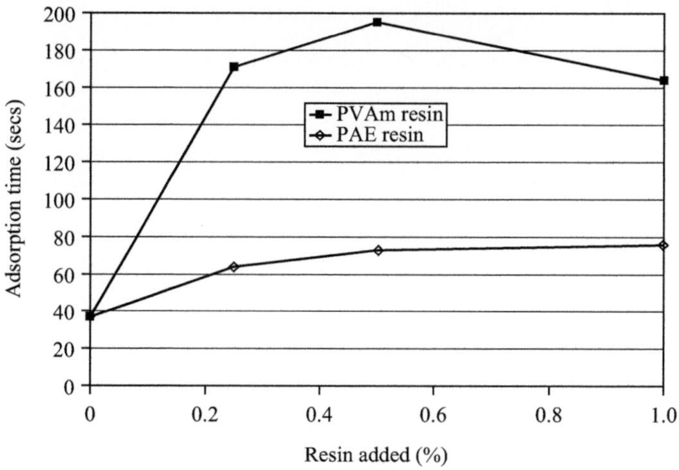

Fig. 8.10 Effect of PVAM wet strength resins on paper absorbency

The wet strength imparted by PVAm resins is permanent, but the dosage levels are low enough that the paper is readily re-pulped. PVAm resins impart lower levels of wet strength than PAE resins at the same dosage level. PVAm resins are very reactive; work over a broad pH range; and are much less sensitive than GPAM or PAE resins to the affects of sulphites, pH, alkalinity and conductivity.

The mechanisms by which PVAm resins impart wet strength to paper are not clear. PVAm resins are amine-containing polymers that are not self-crosslinking resins and, apparently, lack functionality to covalently bond with hydroxyl or carboxylate groups on pulp fibre. Many proposals for the PVAm resins' wet strength mechanisms exist in the literature, including multiple hydrogen bonding, multiple ionic bonding, low temperature amide formation and aminal formation with the cellulose chain ends.

Paper manufactured using PVAm wet strength resins is less absorbent than paper manufactured using PAE wet strength resins (Fig. 8.10). This tends to preclude the use of PVAm resins in the manufacture of hygiene grades.

Chapter 9
Starch in the Wet-End

Peter de Clerck

9.1 Introduction

Starch has been used in papermaking almost since the invention of paper. The global paper industry consumes almost 5 million tonnes of starch per year, making starch the third most important raw material in papermaking. Roughly 20% of this is used in the wet-end.

Cationic starch is the major wet-end starch. The phenomena involved in starch adsorption and the selection of cationic starch need careful consideration, as do the proper preparation and application of starch. The criteria for the use of native, anionic and amphoteric starches are also considered.

An additional source of starch in the wet-end is returned size press starch, which can have a significant influence on wet-end operations and machine runnability.

9.2 Chemistry of Starch

Starch is produced by all plants except fungi. Green leaves of plants contain chlorophyll, which is able to absorb sunlight and utilise the energy to catalyse the formation of glucose and oxygen from carbon dioxide and water; this process is known as photosynthesis. This glucose is transported in the sap and later polymerised into starch (a food reserve) or cellulose.

Note the numbering of the carbon atoms on the glucose unit. There are three reactive sites available for chemical substitution, i.e. positions 2, 3 and 6. The degree of substitution (DS) refers to the number of reacted sites per glucose unit. If all three sites are reacted, then the degree of substitution is three (DS = 3.0).

P. de Clerck
Regional Marketing Manager Pulp and Paper Chemicals, Connell Bros. Co. Ltd.,
345 California Street, 27th Floor, San Francisco, USA CA94104;
Connell Bros. Co. (Thailand) Ltd., B.B. Building, 18th Floor, Unit 18–02\No.54, Sukhumvit,
21 Road (Asoke), Klongyoey-Nuea, Wattana, Bangkok 10110, Thailand
e-mail: pdeclerck@connellthai.com

I. Thorn and C.O. Au (eds.), *Applications of Wet-End Paper Chemistry,*
DOI 10.1007/978-1-4020-6038-0_9, © Springer Science+Business Media B.V. 2009

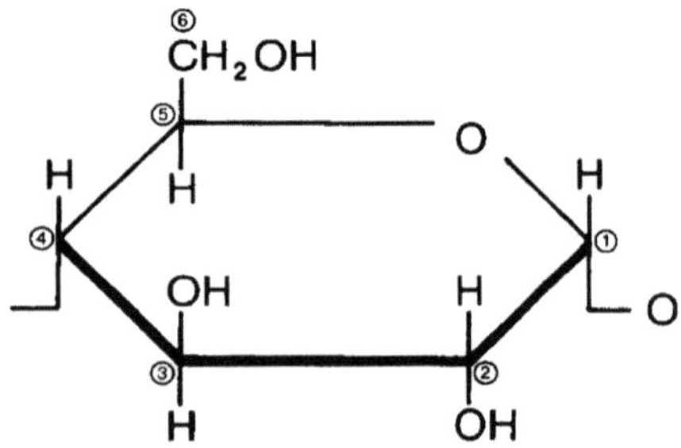

Fig. 9.1 Glucose unit

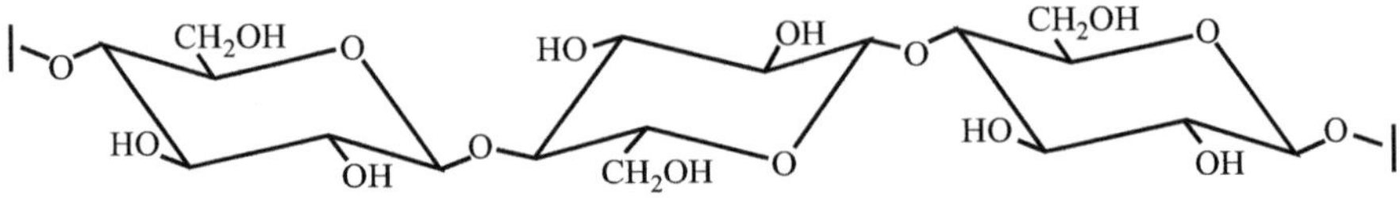

Fig. 9.2 Cellulose

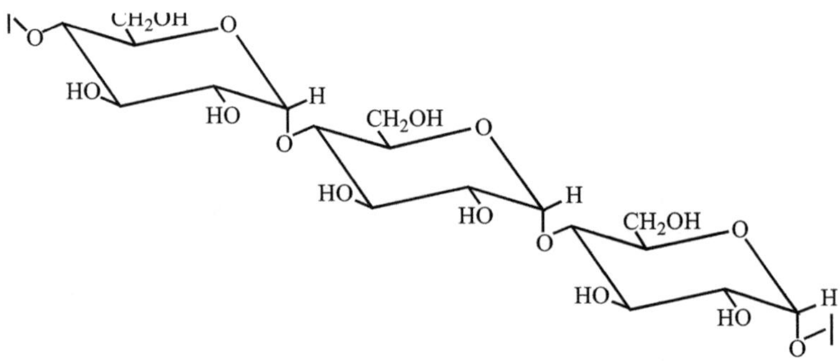

Fig. 9.3 Amylose

Glucose rings may polymerise to give the following materials:

Cellulose – a β *1–4* polymerisation – successive glucose units are inverted, resulting in a polymer chain of up to 2,000–5,000 glucose units

Amylose – an α *1–4* polymeristion – successive glucose units polymerise in a linear 'stepped' conformation, resulting in a polymer chain of 1,000–4,000 glucose units

Amylopectin – Reactions may also take place to form 1, 6 linkages resulting in branched structures. Amylopectin is one of the largest naturally occurring polymers and may contain up to 4,000,000 glucose units.

Starch and cellulose are isomers with similar surface chemistry. There is a significant affinity between these materials. The hydrogen bonding found between hydrated

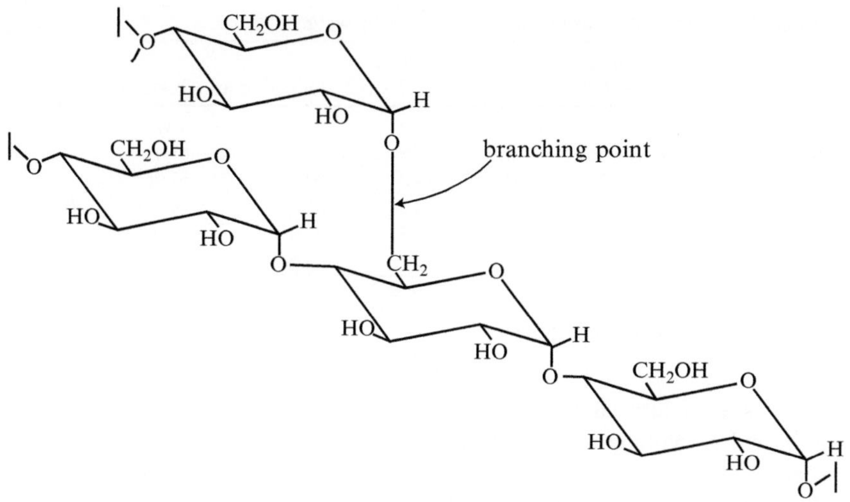

Fig. 9.4 Amylopectin

Table 9.1 Composition of native starches [1]

Starch	Amylose	Amylopectin	Amylose DP	Amylopectin DP
Potato	21	79	3,000	2,000,000
Maize	28	72	800	2,000,000
Wheat	28	72	800	2,000,000
Tapioca	17	83	3,000	2,000,000
Waxy maize	0	100	–	2,000,000

fibre surfaces also exists between hydrated starch surfaces and between starch and cellulose. The non-electrostatic interaction between starch and cellulose plays a critical role in starch performance, e.g. a greater affinity for fully bleached chemical pulp, c.f. a mechanical furnish.

The amylose/amylopectin ratios and amylose DP differ for different starches.

Determination of molecular weight of amylopectin is limited by the gel permeation chromatography columns available. Recent research indicates significant differences in amylopectin DP, with potato amylopectin being by far the largest.

As noted there are major differences between the root and tuber starches (potato and tapioca) and the cereal starches. Waxy starches are special varieties containing little or no amylose, e.g. waxy maize. Substances other than amylose and amylopectin are also present in starch, notably proteins, phosphorous, fats and lipids.

Table 9.2 Composition of raw materials (in % by weight) [1]

Starch	Moisture at 65% RH, 20°C	Lipids	Proteins as N × 6.2	Phosphorous	Taste & odour
Potato	19	0.1	0.1	0.08	low
Maize	13	0.8	0.35	0.02	high
Wheat	13	0.9	0.4	0.06	high
Tapioca	13	0.1	0.1	0.01	very low
Waxy maize	13	0.2	0.25	0.01	medium
Waxy potato	19	0.1	0.1	0.08	low

9.2.1 Phosphorus Content

Phosphorus structures and contents in starches vary with the botanical source, maturity and growing conditions of the plant. Most normal cereal starches contain phosphorus in the form of phospholipids, whereas phosphorus in root and tuber starches is in the form of starch phosphate monoesters [2].

Potato starch is the only commercial starch containing an appreciable amount of covalently bonded phosphate monoester groups, corresponding to a degree of substitution (DS) of about 0.003.

Negatively charged phosphate groups impart a polyelectrolyte character to potato starch and contribute to a low pasting temperature, a rapid hydration and swelling, high water-binding ability and viscosity of potato starch pastes.

9.3 Preparation of Starch

The starch preparation system chosen depends on the mill starch consumption and the level of automation required. Systems using bulk powder, semi-bulk big bags and paper sacks of starch are available. The preparation of a starch solution can be divided into two steps:

- Slurry preparation
- Starch cooking

Preparation of starch slurry can be either by a batch or a continuous system. A continuous system is preferred because the holding time for the starch as slurry is minimised and the risk of microbiological problems is greatly reduced.

9.3.1 Batch Cooking

A basic batch cooker is a tank equipped with an agitator and a steam distribution pipe to heat the slurry to the required temperature. Batch cooking is normally used when a papermill's starch consumption is small.

While heating starch slurry, water is absorbed into the starch granules and bonds between the starch molecules in the granule are broken, causing the

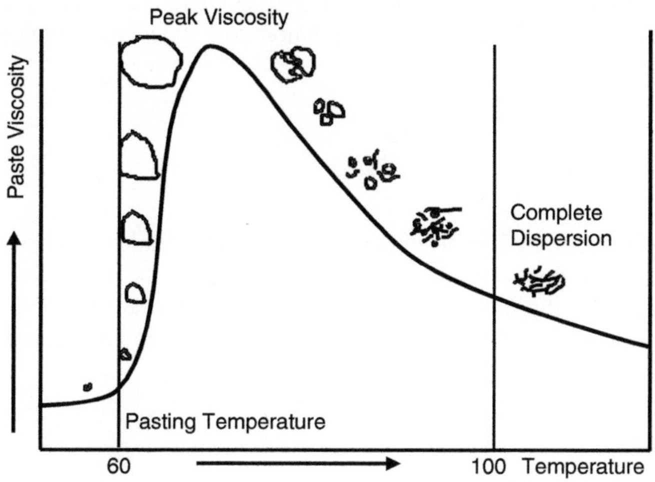

Fig. 9.5 Starch gelatinisation

granules to swell. The critical temperature at which swelling occurs is called the gelatinisation or pasting temperature (about 55–80°C, depending on the type of starch). As heating continues, amylose leaches out and the starch becomes sticky. The paste viscosity increases to a maximum that corresponds with the largest hydrated swollen volume of the granules. Thereafter, the swollen starch granules progressively disintegrate into swollen starch aggregates, eventually achieving a complete molecular dispersion.

The gelatinisation process of starch slurry cannot be defined to take place at a certain temperature but rather occurs over a certain temperature range, depending on the size of the granules.

Complete solubilisation of all starch present does not normally occur unless the starch paste is prepared at temperatures of 100–160°C (depending upon the type of starch). Potato and tapioca can be cooked to a completely dissolved state at about 100–130°C, maize starch at about 130–150°C.

After cooking, the starch paste may be diluted to the desired storage concentration. A biocide should be used to minimise risk for microbiological breakdown of starch. Storage at 60–70°C temperature is recommended [1].

9.3.2 *Continuous Cooking*

Jet-cooking is a continuous method of pasting starch in which steam under pressure mixes completely with starch slurry and rapidly heats it and cooks it within a few seconds. Jet cookers utilise direct steam injection. The jet cooker may be automatically controlled by the level in the cooked starch storage tank.

Jet cookers give a more uniform product as starch granules are completely fragmented and dispersed when cooked at high temperatures (>100°C) and at high pressure. Jet cookers are also more easily automated and a more economical use of raw material is possible.

9.3.3 *Batch Versus Jet Cooking*

The method of cooking has an impact on the starch solution viscosity. The more energy put into the cook, be it thermal energy or mechanical, the greater will be the disentanglement of the starch molecules and hence the lower the solution viscosity. This lower viscosity is due to better dispersion and not to any degradation of the starch molecules.

9.3.4 *Overcooking*

Overcooking has connotations of thermal degradation of starch; such degradation does not occur below 140°C.

For food applications it is desirable to only partially cook starch so that some texture remains. In food terms, a fully dispersed starch is 'overcooked'. For papermaking, only fully dispersed starches will release their full bonding power. Partial cooking of ionic starches will release only part of the starch charge, with resulting variability and difficulties in wet-end charge control. Partially cooked starches may be referred to as 'undercooked' in papermaking applications.

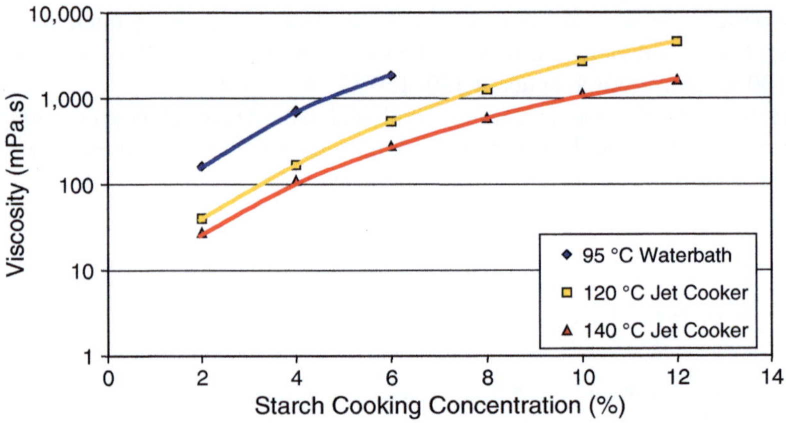

Fig. 9.6 Effect of cooking (cationic tapioca starch, DS 0.030)

9.3.5 Starch Distribution

High-shear pumps will reduce the molecular weight of starch and thus its efficiency. Progressive cavity positive displacement pumps are recommended. Wet-end starch solutions are generally stored at 4% concentration, being reduced to below 1% concentration at the point of addition for fine paper and below 2% concentration for board and industrial grades.

9.3.6 Retrogradation or Set-Back

If a dilute starch solution stands for a prolonged time it gradually becomes cloudy and eventually deposits as an insoluble white precipitate. If more concentrated starch dispersion is allowed to cool it will rapidly set to an elastic gel. Both these are processes of retrogradation, whereby the starch molecules go from a dissolved and dissociated state to an associated condition. The mechanisms of retrogradation are schematically shown in Fig. 9.7 [1].

Retrogradation of starch pastes or starch solutions may produce the following effects:

- Increase in viscosity
- Development of opacity and turbidity
- Formation of insoluble 'skins' on hot pastes
- Precipitation of insoluble starch particles
- Formation of gels
- Syneresis of water from the paste (weeping)

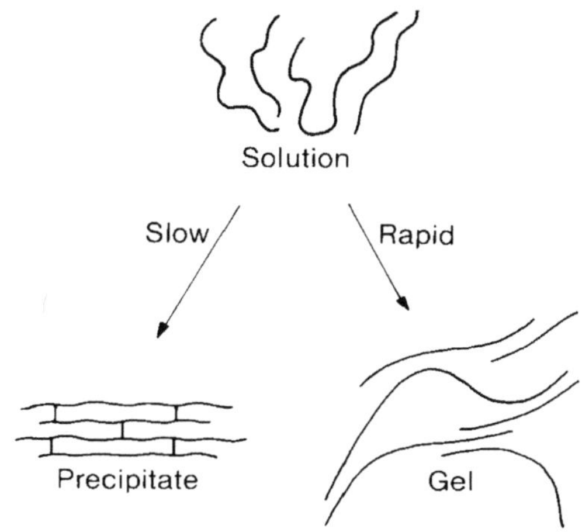

Fig. 9.7 Mechanism of amylose retrogradation

Retrogradation is a complicated process and depends on many factors such as starch type, starch concentration, cooking procedure, temperature, storing time, pH, cooling procedure and the presence of other compounds – notably lipids and other aliphatic hydrocarbons. The retrogradation of starch dispersions is favoured by low temperatures and by high starch concentrations. The retrogradation rate is fastest at pH 5.0–7.0, decreasing at both higher and lower pH.

9.4 Application of Starch

9.4.1 Native Starch in the Wet-End – Starch Spraying

Uncooked starch slurries are sometimes added at the wet-end, intending that the starch cooks in the dryer section. This procedure is satisfactory only if the temperature of the web is high enough to gelatinise the starch granules in situ, otherwise no value is derived from the uncooked starch granules; they merely act as filler and lower the strength of the paper [3].

Uncooked starch is occasionally sprayed between the plies on multi-ply paper and board mills as the simplest solution to the problem of poor or variable inter-ply bonding.

For sprayed starch the retention is by mechanical filtration alone, so a large particle size is favoured. The average particle sizes of potato and tapioca starches are far larger than for maize starch as shown in Table 9.3.

Table 9.3 Average particle size

Starch type	Weight-ave. diameter	Gel Temp.
Potato	40	60–65
Tapioca	25	65–70
Maize	15	75–80

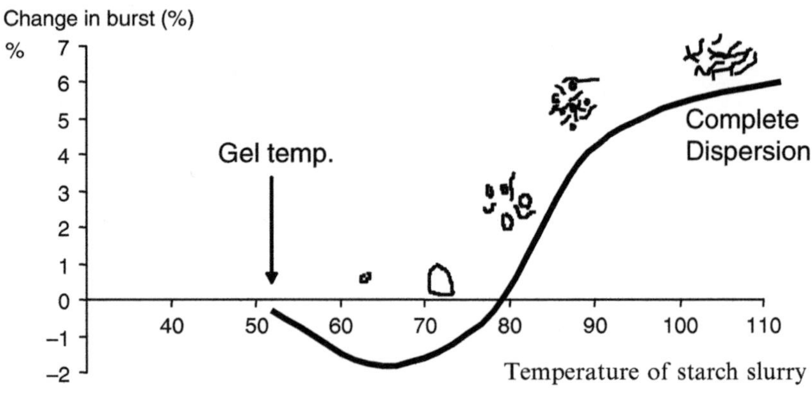

Fig. 9.8 Effect of starch dispersion on strength

Once good retention of the starch is achieved it then becomes necessary to develop its strength/bonding potential by cooking the starch within the sheet. For a starch to cook it needs temperature, moisture and time. The period in which all the conditions required for starch gelatinisation exist within the web is limited to the early part of the main dryer section.

The gelatinisation temperature of the starch is very significant in determining the onset of starch cooking in the dryer section and hence the time available to cook the starch before the local moisture content in the paper/board becomes too low for further gelatinisation to occur. This factor becomes more critical as machine speeds and pressing efficiencies increase, further reducing the time and moisture available for starch gelatinisation. Starch derivatives with pasting temperatures as low as 45°C have been developed.

Some mills spray starch slurry to the Fourdrinier wire on single-ply webs. The position of the starch spray beam is critical in these applications; spraying after the dry line results in starch sitting on the surface of the sheet with potential problems of felt filling and dryer sticking, while spraying close to the headbox results in the loss of most of the starch with the drainage water. The optimum lies somewhere between these positions and must be found by experimentation.

9.4.2 Anionic Starch in the Wet-End

Anionic starches include carboxymethyl starch and starch phosphate esters. Starch phosphates were first produced commercially in 1960. They were used extensively in Europe in acid systems, being retained in the sheet by alum. Their addition to the papermaking system requires a minimum of 1% alum (dry fibre basis) in the beater stock and a pH in the range of 4.3–6.0 [4]. These starches proved hugely successful in replacing the native (unmodified) beater starches then in use.

The use of anionic starches in the wet-end declined rapidly following the introduction of cationic starches and their use almost ceased with the subsequent move to neutral sizing. There has been a revival of interest in anionic products for use in dual-starch systems [5].

9.4.3 Cationic Starch in the Wet-End

Early cationic starches were tertiary aminoalkyl starches. Tertiary amines require a proton to remain cationic and were suitable for use in acid papermaking systems. At increasing pH levels, the tertiary amine eventually looses its charge. Quaternary amines remaining cationic throughout the papermaking pH range and now represent the majority of cationic starch types.

9.4.4 Cationic Starch Production Process

Wet process products are made batch-wise with starch slurries of about 35%. Reagents are added and the reaction proceeds under controlled conditions for several hours. The reaction is then stopped and the product washed and dried.

For the semi-dry process, dry starch is fluidised in a special mixer to which are added the various reagents in dilute form. Water needs to be present to carry the chemicals into the starch granule. The process is continuous with the product being dried in line, sieved and bagged.

For fuller details of the processes, please see Swinkels' Industrial Starch Chemistry [1].

Starch samples were ground down and DS was analysed at varying stages of grinding by van de Steeg [6]. It was found that for wet process products the DS was constant at all levels of grinding, while for semi-dry product the DS was reduced at a certain stage. This was seen as evidence that the DS is not even throughout the granule in the later process. Note that the average DS is the same in both cases.

The difference lies in the distribution of DS, the distribution being broader in the semi-dry product.

The more closely specified wet process product would be suited to closely defined, ordered systems such as pulp-based fine paper machines. Wet-end charge balancing could be carefully controlled.

The broader charge distribution of the semi-dry product has an advantage in 'dirty' systems, where there exists a variety of anionic species of differing charge affinities – such as waste paper systems and de-inked pulps where washing is not fully optimised. It appears that the higher DS fraction is first exhausted by the more aggressive anionic species, leaving the remainder to react with the fibre.

Replacing a wet process product with the equivalent semi-dry product in fine paper machines resulted in increased flocculation in a number of trials, necessitating a reduction in the starch or polymer addition in order to regain formation.

Semi-dry products are more difficult to disperse than wet process products and generally require a jet cooker.

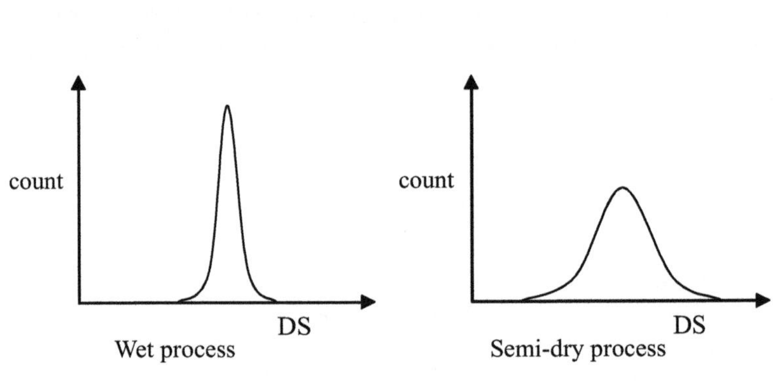

Fig. 9.9 DS distribution upon grinding starch

9.4.5 Cationic Starch Adsorption

When a polyelectrolyte adsorbs on an oppositely charged surface the adsorbed amount compensates, or slightly over-compensates, the surface charge [6]. Adsorption of cationic polymers continues even after the zeta potential is reversed [7]. This is the case for cationic starch.

Cationic starch will be adsorbed by fibre until either of the following takes place:

1. The fibre surface is fully covered by starch.
2. The fibre surface charge has been fully (over) compensated.

Polyelectrolytes with molecular weight between 5,000 and 10,000 are able to enter most fibre pores, but starch molecules are too large to penetrate into the fibres. So only the surface charge on the fibres is available for interaction with cationic starch. The accessible fibre surface for cationic starch (the hydrodynamic surface) is estimated to be about 6 m^2/g, which is only 10% of the combined hydrodynamic and internal surface area accessible to small ions [6]. Papermaking fibres become surface saturated at 16–25 kg/t starch on fibre [8]. For large polymers such as starch, adsorption increases linearly with hydrodynamic surface area.

9.4.6 Effects of Salts

Polyelectrolyte adsorption is a subtle balance of electrostatic and short-range non-electrostatic interactions. For a given combination of charge density, segment charge and non-electrostatic interactions, this balance can be affected by changing the concentration of simple electrolytes [6].

Starch adsorption decreases with increasing specific conductivity and is sensitive to the type of salt cation. The bivalent cations Ca^{2+} and Mg^{2+} appear to be ten times as effective as Na^+ in suppressing the adsorption of cationic polymers.

Polyelectrolytes desorb if the salt cations are more effective in compensating the charge of the adsorbing surface [6]. Calcium ions are particularly effective at

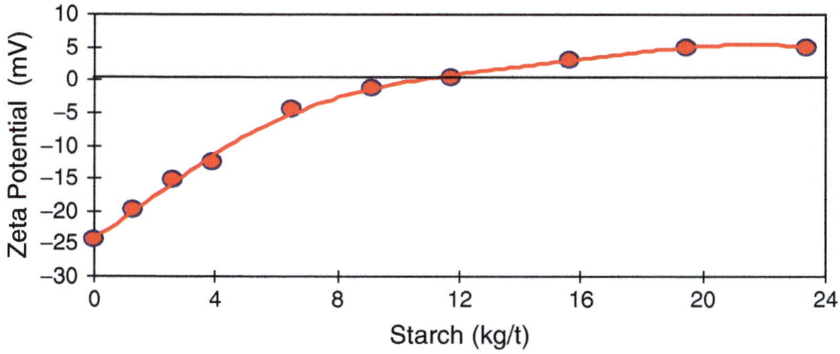

Fig. 9.10 Cationic starch adsorption isotherm on fibre

displacing adsorbed polyelectrolytes from cellulose. The larger effect of the divalent cations is caused not only by more effective screening, but also by specific interactions between Ca^{2+} or Mg^{2+} and carboxylate groups [6]. The level of free calcium ions in papermaking systems has increased notably with the introduction of PCC, gypsum filler and more closed wet-end systems.

Polyelectrolytes with higher charges per segment (DS) adsorb less at low specific conductivity. At a low specific conductivity the starch with the lowest DS is expected to adsorb most, whereas at high specific conductivity the starch with the highest DS will adsorb best [6].

The non-electrostatic interactions between starch and cellulose make starch less susceptible to desorption than other polymers.

9.4.7 *Effect of DS*

The largest effect of DS occurs at the lowest salt concentration, the polyelectrolyte adsorbs until the surface charge is compensated [6]. The amount adsorbed is inversely proportional to the DS.

At higher salt concentrations, the electrostatic attraction between polymer and surface is screened and the adsorbed amount decreases. Thus, adsorption of low DS starch (= ~0.015) decreases rapidly with increasing salt concentration [6].

The screening of the charge by increasing salt concentrations results in a need for higher DS products at higher conductivities.

It is clear that increasing salt concentration leads to shrinking of the cationic starch molecules due to screening of repulsion between the charged groups [6].

Increasing the DS of the starch results in greater segment-segment repulsion and so resisting the compressing effect of the salt. This may be seen in Fig. 9.11.

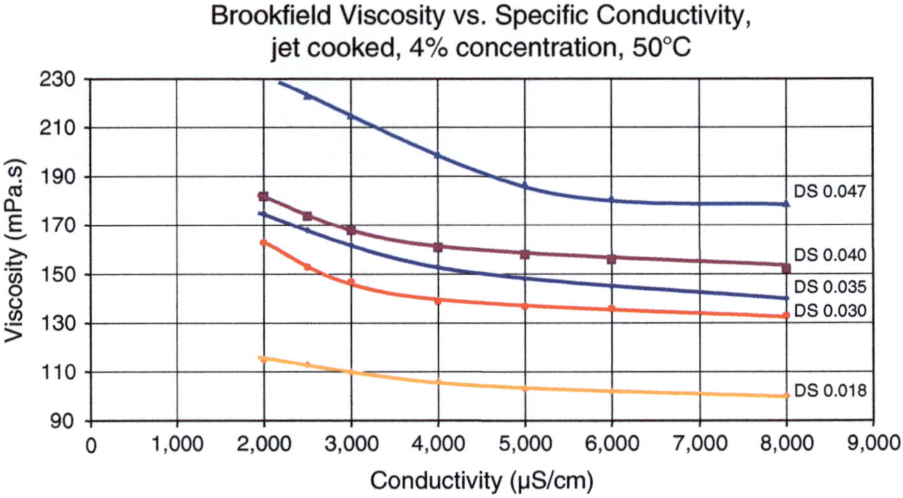

Fig. 9.11 Effect of degree of substitution on salt resistance

9.4.8 *Mixing Shear and Time Effects*

For particles of >10 μ radius, agitation is far more important than diffusion in causing inter-particle collisions. Increasing time and shear will increase the number of collisions and so cause an increase in adsorption [9]. Greater polymer adsorption is achieved when addition is to a zone of high stock turbulence.

Adsorbed polymers will slowly collapse as the loops and tails become less extended. The measured zeta potential of the fibre will fall as the effective radius of charge interaction collapses. Hydrogen bonds form between adsorbed starch molecules and fibre surfaces

From DDJ results it was concluded that the retention of a cationic polymer in a paper web will be maximised by allowing sufficient time and agitation for the dissolved polymer molecules to contact and become adsorbed on the surfaces of fines and fibres. Such conditions favour the use of these polymers as dry strength additives. The reverse is true in regard to their use as retention and drainage aids, i.e. thick stock addition for strength, thin-stock for retention & drainage [9].

The final factor which must be considered is the ability of the aggregate to withstand external forces. An increase in agitation is desirable as this will increase the collision frequency and hence the rate of reaction but at higher shear dispersion will predominate as the hydrodynamic forces increasingly tend to strip the additive off the fibre [10]. The number of initial attachment points to the fibre (and hence the attachment strength) is a function of DS. High drainage shear – such as encountered on modern gap formers – requires a higher starch DS than a Fourdrinier making the same paper grade.

9.4.9 *Amphoteric Starches*

Amphoteric wet-end starches carry both cationic and anionic substitutions.

Cationic potato starch was found to be significantly superior in performance to other starches in systems containing alum. Vander Burgh [11] claimed that in alum-containing papers cationic potato starch was twice as effective as cationic

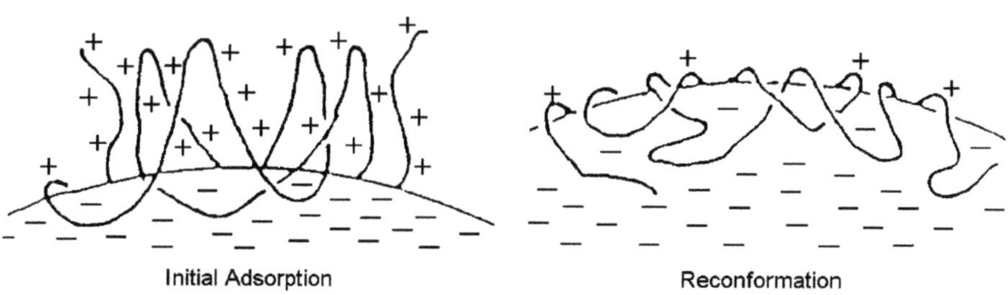

Fig. 9.12 Reconformation of an adsorbed polymer

maize starch. Harvey [12] showed that cationic potato starch and amphoteric maize starch were equally well retained under those circumstances.

Due to the natural phosphate content present in the amylopectin fraction, potato starch has an anionic character. A cationic potato starch is actually a blend of cationic amylose and amphoteric amylopectin.

Attempts were made to replicate potato starch performance with other starches and in 1965 National Starch filed their landmark patent application for amphoteric starch.

Early amphoteric starches contained tertiary aminoethyl ether and phosphate ester groups in which the preferred molar ratios of anionic to cationic groups was from 0.07 to 0.18. Later patents cited phosphate ester and quaternary amine with ratios of 0.014 to 0.92. These derivatives provided synergistic effects and were effective filler retention aids and paper strengthening agents over the pH range from 4.0 to 10.0 [4].

Very early in the use of cationic starches, it was recognised that acidity from alum in moderate to high levels can change the requirements for a cationic starch to those of an amphoteric starch. The highly cationic alum at low pH competes for fibre and filler adsorption sites with the cationic starch [13]. It is postulated that interaction between the alum and the anionic phosphate affects starch retention. The anionic charge does not appear to provide an appreciable benefit in alkaline systems [14].

The presence of alum is usually required for a successful result using amphoteric starch. Alum exhibits its maximum cationicity around pH 5.0 – corresponding to the maximum retention of amphoteric starch in Fig. 9.13.

Charges on the amylopectin chains are screened in high conductivity systems. For a cationic amylopectin molecule this will lead to a reduction in the repulsion forces between the cationic sites, leading to a compression of the molecular radius, lower viscosity and a lesser effective bonding radius. For an amphoteric amylopectin molecule,

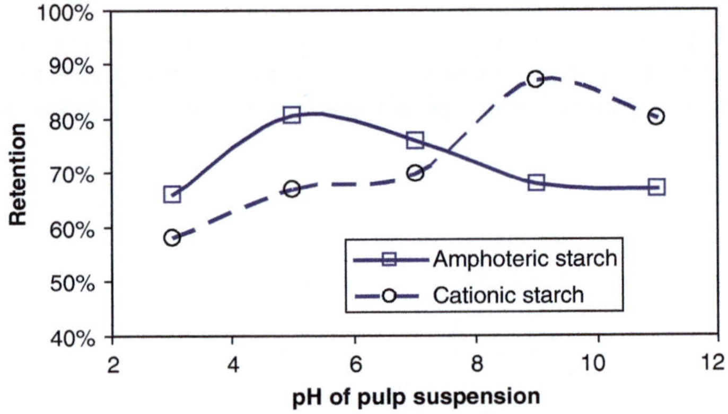

Fig. 9.13 Retention of cationic and amphoteric starches in handsheets (addition level of starch derivatives is 2% on dry weight of pulp) [15]

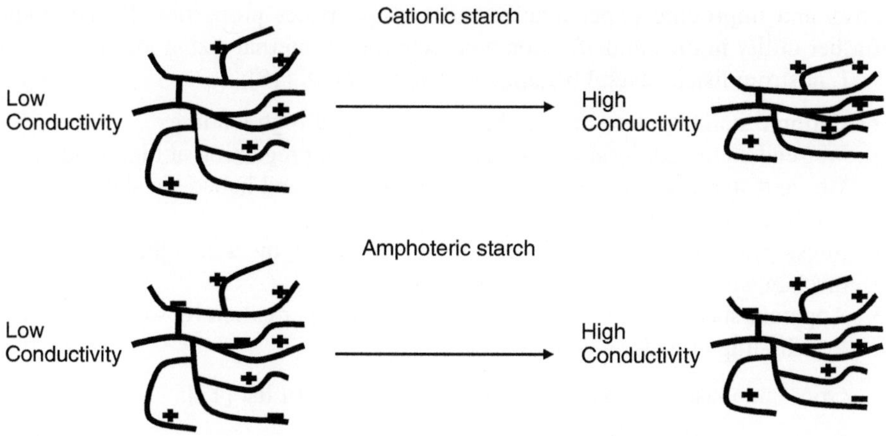

Fig. 9.14 Effect of increased conductivity on cationic and amphoteric starch

high conductivity will screen the attractive charges between the oppositely charged sites and may allow the molecule to expand.

This resistance to compression may also be gained by increasing the degree of substitution of a cationic starch, as shown in Fig. 9.11.

It is claimed that an amphoteric starch is more tolerant to changes in anionic or cationic materials in the system, acting as a charge buffer. It is also claimed that the anionic group on the starch complexes the soluble cationic materials and reduces system sensitivity to changes in pH and zeta potentials [16].

9.4.10 Dual Starch Systems

The sequential use of anionic and cationic wet-end starch was described as early as 1975. In an acid furnish, anionic starch would be added first, adsorbed onto the fibre by alum, and then a cationic starch was added. A reversed addition sequence has been proposed for alkaline papermaking systems [8]. There are numerous patents covering sequential additions of oppositely charged starches [17, 18].

9.5 Function of Starch

9.5.1 Effects of Starch in Paper

The initial use of cationic starch was for strength and retention. Cationic starch was found useful in improving machine run ability and efficiency, lowering furnish costs and improving paper quality and end-use paper properties. It later found further utility in the emulsification and retention of internal sizing agents.

Lindström listed several benefits of cationic starch [18]:

- Cationic starch gives a denser sheet, especially at high loadings.
- Wet-end starch addition is more effective than impregnation at high loadings.
- Wet-end starch greatly increases the drying stress and hence the sheet modulus – stiffness improves.
- Stress concentration is reduced – it is proposed by a mucilage lubrication mechanism.
- Apparent sheet density is increased to a much larger extent by wet-end addition than by impregnation addition.

Grief and Gasper produced a more extensive benefit list [13]:

- Better runnability and fewer wet-end breaks
- Improved filler retention and better filler distribution
- Filler retained by cationic starch is also bonded in the web
- Reduced linting and dusting
- Reduced hardwood vessel picking
- Reduced pick outs
- Stronger surface fibre bonding in the base sheet, which increases the interface adhesion of the coating to the base paper
- Internal bond (Scott Bond) increases
- Better print fidelity and sharper images attained on offset (both sheet and web fed), letterpress, and rotogravure printing presses
- On bag paper, tear increases have been realised when refining is altered and cationic starch is used to maintain tensile and Mullen with less refining. A more 'squared' sheet is obtained with greater equalisation of machine and cross-machine tensile.

9.5.2 Strength Mechanisms

The prime purpose of adding starch to paper is normally for strength improvement.

The development of inter-fibre bonding is perhaps the most fundamental issue of papermaking. Such bonding is due to short-range hydrogen bonding, which arises as the carboxyl (COOH) groups on cellulose ionise and associate with the dipole on water molecules, i.e. the fibre surface is hydrated. During drying the water 'bridges' between the hydrated fibre surfaces become shorter and surface tension draws the surfaces together. Eventually only inter-fibre bonds remain.

Hydrogen bonding can only develop over distances of less than 3 angströms (0.3 nm). Rough fibre surfaces have asperities larger than that, thus physically preventing the formation of hydrogen bonds.

Beating produces fibrillation and hydration and some hemicellulose is also leached from the fibre, resulting in a more gelatinous fibre surface. This more gelatinous surface allows greater bonded areas to develop.

Adsorbed hydrocolloids (starch, gums and polymers) may be expected to act in a similar fashion to fill the discontinuities and voids between the fibres.

Cationic starch would be retained on the most active areas of the fibre surface, and thus the most likely potential bonding sites. Irreversible retention occurs at the initial contact, and final strength development, unlike with gum or polyacrylamide, is not dependent on migration of starch to the fibre crossings during drying [6].

Clearly, molecular size will be important as larger molecules may bond over larger distances. The state of dispersion of the starch is therefore an important factor [19].

9.5.3 Strength and Flocculation

Handsheet testing of strength additives often gives results that bear little relation to the results obtained on a paper machine. This is often due to the greatly lower drainage shear and retention obtained in a handsheet former and the subsequent difference in sheet composition and structure. Since most wet-end additives are ionic in character, the longer de-watering time and lower consistency in a handsheet former will often result in much greater charge-induced flocculation than would be found in industrial practice.

Roberts and colleagues [20] conducted an elegant series of experiments in which the breaking length and formation index were measured for different starch additions and contact times. It was concluded that native starch and cationic starch improved paper strength equally at similar formation numbers.

This was a very important debunking of the myth that increased cationicity in starch leads to enhanced paper strength (extension of this logic would be that anionic starches are debonding agents).

For undegraded starches it is the amount of starch retained in the sheet that determines the strengthening effect. The cationicity of the starch serves to increase the starch retention; too high a cationicity can cause strength loss through formation changes.

9.5.4 Filled Papers

The strengthening effect of cationic starch is much greater in a filled sheet and for very high filler levels; wet-end addition is greatly superior to surface sizing for strength improvements. When cationic starch is added to highly filled papers the sheet density increases notably, indicating a significantly higher level of bond formation. Laboratory sheets containing 85% filler and 4.5% cationic potato starch still exhibit 'rattle', whereas the 'paperiness' of the sheet falls rapidly above 30% filler when cationic starch is not used.

In filled sheets it has been proposed that part of the strengthening effect of the starch is its action as a mucilage or lubricant in the early draws on the paper machine. A filler particle acts as a discontinuity between two fibres in the sheet and

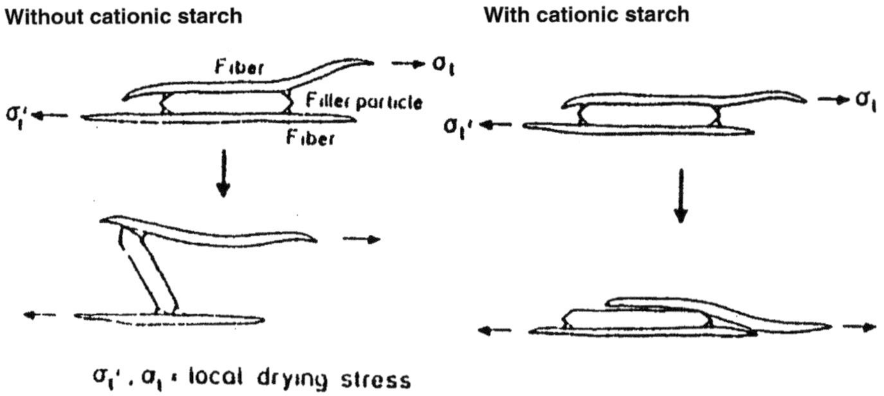

Fig. 9.15 Stress release in the presence of cationic starch [21]

so acts as a focus for stress concentration. During drying and in machine draws, local stresses are set up in the sheet. If the local friction between the filler particle and the fibre surface is large, then the fibre network locks until the drying stress exceeds the local failure stress. When local failure occurs the water meniscus between the fibres may be broken; the fibre surfaces are then too far apart for hydrogen bonding to develop and reconsolidation of the bond is unlikely.

If the filler particle is surrounded by mucilage such as cationic starch, the local friction between the filler particle and the fibre is reduced and 'slipping' may occur under the action of localised drying stresses. When such 'slipping' occurs, local stress concentrations are decreased and breaking of water menisci does not occur. Hence bonding and sheet consolidation can continue.

From the above figure it is proposed that in the absence of cationic starch more bonds may be 'unhinged' in a filled sheet. This would result in better opacity and bulk with lower sheet strength – characteristics normally found in such sheets.

9.5.5 Retention

The introduction of cationic starch gave significant retention improvements over the use of alum. The use of cationic starch as a retention aid was eclipsed by the introduction of polyacrylamides and other polymers, being consigned mainly to the role of a strengthening agent. The advent of micro-particle retention systems revived the position of cationic starch as a key retention system element – particularly in silica nano-particle systems. (These are covered elsewhere in this book.)

9.5.6 Size Emulsification

Cationic starch also has a major role in the emulsification and retention of highly reactive internal sizing agents. These emulsions have no charge other than the charge provided by the emulsifying agent – size retention is dependant on the encapsulating starch for its attraction and fixation on the fibre surface. For this purpose it is normally recommended to use a higher DS starch than would be recommended if using starch solely as a strength additive. Cationic DS 0.027–0.045 is normally used for size emulsification. Special starches with low pH or enhanced filming properties may also be required. Amphoteric and cross-linked starches are not suited for this application.

Starch added as an emulsification agent makes little or no contribution to inter-fibre bonding. When a size emulsion droplet is retained on a fibre surface, upon rupture of the droplet the encapsulating starch attached to the fibre surface remains covered with the hydrophobic size and is unable to contribute to fibre bonding. Such starch should not be included when considering the starch requirements for increased paper strength.

9.5.7 Anionic Trash Catcher

Anionic colloids may complex with cationic starch, reducing the ability of the starch to bond to the fibre surface. Hydrocolloid complexes may also be formed, resulting in poor de-watering and pressing efficiencies. Disturbing anionic trash should be neutralised with a suitable anionic trash catcher before addition of cationic starch to the system. This topic is well covered in several works by Pieter Brower [22–27].

Anionic trash catchers are either inorganic salts (aluminium, zirconium) or high charged cationic polymers. Extremely high DS, low molecular weight cationic starches are available for use in this application and are found to quench OBAs much less than polymers such as poly-DADMAC or polyacrylamide.

9.6 Mill Experience with Starch

9.6.1 Wet-End Strategy and Starch Selection

For a charge-balanced approach to the wet-end, it is recommended to achieve a final zeta potential close to zero for all wet-end components. This position should normally be achieved by approaching the neutralisation point in progressive manner starting with the anionic fibre – swinging from anionic to cationic and back is not normally recommended. As cationic starch is a major source of cationic charge to the system, it is necessary that the cationicity of the starch solution is constant if system stability is to be obtained. Considering starch cooking and from Fig. 9.5, it is apparent that only a fully dispersed starch will release its full cationicity into the system, with variations in the cook quality resulting in variable charge release. In charge-balance systems it is essential that fully cooked starches are used. If a partially cooked starch is used it should be an amphoteric starch with a net charge close to zero.

The development of micro-particle retention systems radically changes the chemistry. In these systems an overdose of cationic material is added to the thin stock, such that the system becomes cationically dispersed. An anionic micro-particle is later added to flocculate the stock. Under these circumstances a complete starch cook out may not be necessary – especially if the system is being over-charged with a polymer rather than starch. Cross-linked cationic starches that do not fully disperse may be used in such systems. The partially dispersed starch acts partly as a colloid and partly as a particle, having a larger active radius than a fully dispersed starch molecule. Benefits in starch retention are claimed.

9.6.2 Starch Hygiene

Any consideration of starch must include starch hygiene. Starch is a major food source and virtually every living organism can generate enzymes to break down starch.

Starch breakdown can have dramatic consequences for paper machine operation. The following is a case study reported in the UK [28].

Major retention swings were noted on a paper machine. Flow meters detected no corresponding changes in retention aid or cationic starch addition. The FPR was cycling in a 40-min cycle, corresponding to the batch cooking cycle of the cationic starch.

The cause was found to be microbiological contamination in the starch tank. Whenever a fresh batch of starch was pumped over, the tank viscosity and FPR rose. Over time, the biological activity in the starch tank reduced the viscosity of the starch and the FPR fell. The cycle was repeated when the next batch of starch was pumped over.

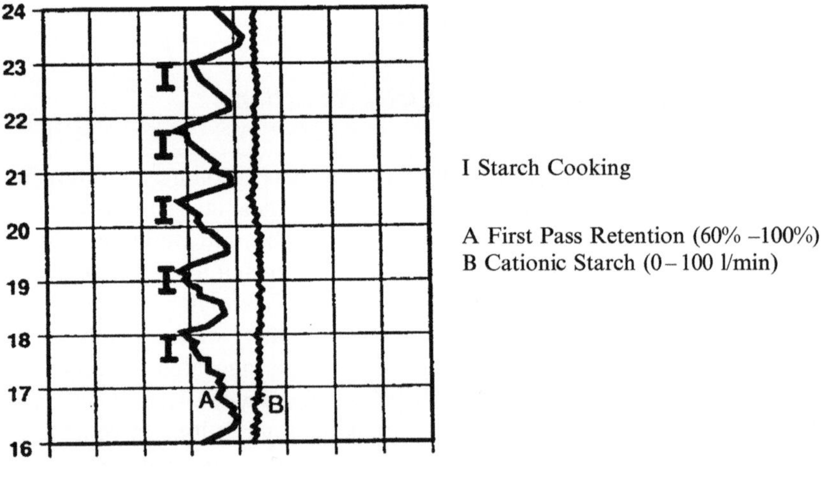

Fig. 9.16 FPR cycling and starch cooking

The starch tank was not emptied during the run cycle and so provided a continuous reserve of micro-organisms. On adding an appropriate biocide to the starch tank a constant, high FPR was achieved.

This case serves to demonstrate both the need for a biocide programme in any starch system, and the benefit of using high molecular weight undegraded wet-end starch.

9.6.3 Returned Size Press Starch

The best approach to slime control is to maximise the wire retention of biodegradable materials, notably starches. This particularly applies to size press starch returned with mill broke. Size press starch can be a source of considerable wet-end problems. Unless retained quickly, size press starch will build up in the circuit waters, contributing to various wet-end problems (deposition, chemical interference and so on), and finally, to waste water loads.

On a fine paper machine, size press starch addition is typically 1.2 gsm per side. For a 70 gsm copier grade this equates to 34 kg/t. Hence if a mill is running 10% broke the returned size press starch amounts to 3.4 kg/t in the furnish. This can have considerable impact on wet-end chemistry. Size press starch is of necessity viscosity-reduced, and so contributes little to the paper strength on recycling. However, it is believed that while the starch chains are too short to bridge between fibres in the forming sheet, it may contribute to an increase in relative bonded area after pressing.

Size press starch may be non-ionic, from in-mill enzymatic conversion or thermo-chemical conversion of native starch. It may be anionic if an oxidised starch is used. (Note this refers to chemically modified oxidised starch. Many American texts wrongly refer to any viscosity-reduced starch – including enzyme-converted and thermo-chemically converted starch – as oxidised starch. Such processes do not form any carboxyl or carbonyl groups on the starch.) Furthermore, cationic and amphoteric starches are frequently applied on multi-purpose office paper for improved ink-jet printability.

A 1,000 t/day fine paper machine changing from a cationic to a non-ionic size press starch therefore removes 3.4 t/day cationic materials from the wet-end. This can lead to a serious charge imbalance in the wet-end.

An example is shown below for a gap former machine that changed from oxidised (anionic) size press starch to enzymatic conversion of native tapioca starch. The graph shows the accumulated lost time due to size press breaks over a 120-day period – 60 days before the change and 60 days after.

Time lost for size press breaks more than doubled. The change in machine run ability was clearly evident in the accumulated lost time chart. The lost time increase at the size press is mainly due to the quality difference between a fully modified starch and an in-mill enzymatic hydrolysis.

More surprising was that lost time for press section breaks more than doubled. The web is fully supported until the final press roll on this machine. There was a

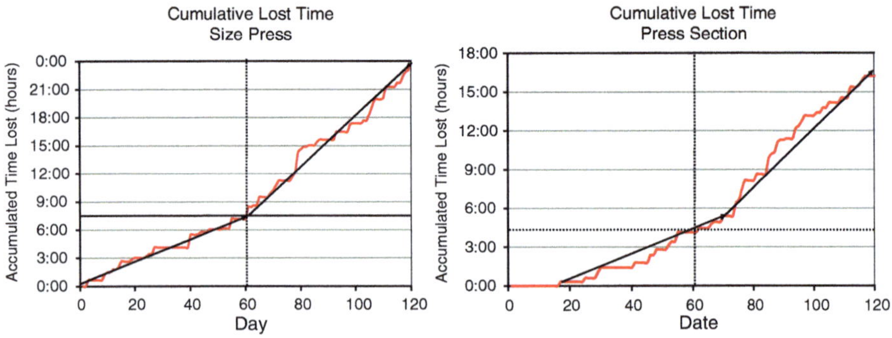

Fig. 9.17 Accumulated lost time due to breaks over 120 days

lag before the full effect kicked in, this being due to the time cycle for exhausting the broke returns from the associated cut-size converting operation.

The dramatic change in runnability was due to the change in wet-end chemistry. When running oxidised (anionic) surface size, the returned starch in the broke stream was fixed in the sheet with PAC. No free starch was evident in the wet-end system. On changing to enzyme-converted native starch (non-ionic), the amount of circulating hydrocolloids rapidly increased as the non-ionic material was poorly retained on the wire. These hydrocolloids had a very high surface area and interfere with adsorption of other chemicals onto the competing cellulose surface. The additional hydrocolloid load entering the presses reduced de-watering, leading to a weaker sheet and increasing the number of press breaks. Mill BOD levels rose accordingly, but the effects on micro-organism levels and biocide demand were not disclosed.

Overall, the increase in lost time was far more costly than the cost savings on the starch.

9.7 Summary

Starches are used on a large scale in the paper industry as wet-end additives. They improve dry strength, retention of fines and fillers, and drainage. They are also essential components of modern retention strategies and sizing programmes.

Closure of the white-water systems in the paper mills has increased the concentration of detrimental substances and the specific conductivity. This has increased the required DS of the starch.

For a wet-end starch to be effective it is important that it is

1. Of sufficient molecular size, i.e. largely undegraded
2. Properly dispersed
3. Properly retained – preferably on the fibre

Proper care must be taken to prevent starch degradation in storage by use of biocides. Degraded starch can severely impact retention, drainage and strength.

Attention must also be given to the wet-end chemistry, the selection of the correct starch addition point and the use of efficient retention aids.

Additions to the stock must be made at 1% solution concentration or less for complete dispersion in the stock stream [19].

Starches for size emulsification need a higher DS to ensure retention of the emulsion droplet. Starch used for size emulsification makes little or no contribution to sheet strength.

Increased drainage shear on high speed gap formers requires the use of higher DS starches.

Acknowledgements Particular reference is made to the following authors: J.J.W. Swinkels for information on native starches and fundamental structure and chemistry, H.G.M. van de Steeg for her dissertation on starch adsorption, P.H. Brower for the development of starch application in the wet-end, Professor T. Lindstorm for fundamental wet-end chemistry.

References

[1] J.J.M. Swinkels (1985), Sources of starch; its chemistry and physics, in G.M.A. van Beynum and J.A. Roels eds., *Starch Conversion Technology*, Marcel Dekker Inc., New York and Basel, pp. 15–46.

[2] Tunyawat Kasemsuwan and Jay-Lin Jane (1996), Quantitative method for the survey of starch phosphate derivatives and starch phospholipids by31P nuclear magnetic resonance spectroscopy, *Cereal Chem.*, 73(6): 702–707.

[3] James P. Casey (ed.) (1980), *Pulp and Paper Chemistry and Chemical Technology,* 3rd edn. Wiley-Interscience.

[4] B. T. Hofreiter, Natural products for wet-end addition

[5] P. H. Brower, P.H (1997), 'Anionische Massestärke rein – Oberflächenstärke raus ?', *Wochenblat fur Paper*, 125 (19): 928.

[6] H.G.M. van de Steeg (1992), Cationic Starches on Cellulose Surfaces: A Study of *Polyelectrolyte Adsorption*, Wageningen.

[7] H. Tanaka, K. Tachiki, and M. Sumimoto (1979), Adsorption of cationic polymers onto bleached kraft pulp: *TAPPI Journal*, January, 62(1).

[8] T.A. Wielema and J. Baas (2005), Wet-end starch and increasing the mineral level in paper, *PIRA conference*, Lisbon.

[9] D. Abson and D.F. Brooks (1985), Wet-end behaviour of dry strength additives, *TAPPI Journal*, 68(1), January.

[10] A.W. McKenzie (1968), The retention of wet-end additives, *Apia Journal*, 21(4), January.

[11] L.F. Vander Burgh (1970), Use of cationic starch in papermaking. *Pulp and Paper Magazine Canada*, 7 1 (18): 81–84.

[12] R.D. Harvey (1985), Retention of cationic starches, *TAPPI Journal*, 68(3): 76–80.

[13] D.S. Greif and L.A. Gasper, Cationic starch as a wet-end additive, *Dry Strength Additives*, TAPPI Press, p. 95.

[14] R.D. Harvey (1985), Retention of cationic starches, *TAPPI Journal*, 68(3).

[15] J. Yoshizawa, A. Isogai and F. Onabe(1998), Cationic and amphoteric starches on handsheets, *Journal of Pulp and Paper Science*, 24(7), July.

[16] B.M. Jensen and J.C. Winters (1986), Cationic potato starches proven effective as wet-end additives, *Pulp & Paper*, April.

[17] US Patent 20020104634 – Starch compositions and the use thereof in the wet-end of paper preparation.
[18] T. Lindstrom, P. Kolseth and P. Naslund, The dry strengthening effect of cationic starch wet-end addition on filled papers, P589, *FR3285 Papermaking Raw Materials*
[19] R.W. Radley (ed.) (1976), *Industrial Uses of Starch and Its Derivatives*, Springer, New York.
[20] J.C. Roberts, C.O. Au, G.A. Clay and C. Lough (1989), Effect of C^{14} labelled starches on dry strength, the effect of C^{14} labelled cationic and native starches on dry strength and formation; P88, *TAPPI Journal*, October.
[21] H. Nanko and J. Ohsawa (1989), *Fundamentals of Papermaking*, 9th Fund. Res. Symp., Cambridge p. 783.
[22] P.H. Brower, P.H., M.A. Johnson and R.H. Olsen (1998), Starch and retention, in J.M.Gess (ed.) *Retention of Fines and Fillers During Papermaking*, chapter 10, TAPPI Press, Atlanta, GA.
[23] Brower, P. H. (1989), *Woch. for Paper*, 117(19): 81.
[24] Brower, P.H (1991), *TAPPI Journal*, 74(4): 170.
[25] Brower, P.H., 1992 TAPPI Papermakers Conference Proceedings, TAPPI Press, Atlanta, GA., p. 301.
[26] Brower, P.H. (1996), *Woch. for Paper*, 124(4): 19.
[27] Brower, P.H. (1993), Proceedings of PIRA Conference, 'The Chemistry of Papermaking', PIRA, Leatherhead.
[28] G. Street (1987), Closing the technology gap at the wet-end, *Paper Technology*, November, p. 678.

Peter de Clerck is Paper Division Manager for Connell Brothers Co. Previously he was with AVEBE (Far East) Pte Ltd.

Chapter 10
Paper Colouration

Barry Mclow

10.1 Introduction to Colour in Paper

10.1.1 Colour in Paper

Of all the products added to paper, colour is often seen as the least important as, generally speaking, it adds nothing to the final properties of the paper. However, incorrect use of colour can turn paper, which would meet all the physical properties required of it, into paper which is unsuitable for sale, purely and simply because it is the wrong colour or shade. Correct use of colour, therefore, is of utmost importance to the papermaker.

10.1.2 Why Is Paper Coloured?

In order for colour to be seen, three factors are needed: the object itself, a light source and a receptor. When any of these factors are changed, a different colour will be seen.

The same object will look different when viewed in daylight compared to, say, a tungsten light source, due to the different spectral energy distributions of the two light sources.

When light is shone onto an object, invariably some of that light will be absorbed and some will be reflected, and it is this reflected light which enters the eye and sends a signal, via the optic nerve, which the brain interprets as colour.

Within the eye itself there are receptors known as cones which are able to differentiate between the three primary light sources that enter the eye, namely red, green and blue. Where one or more of these receptors is weak, a person will suffer from colour blindness and be unable to distinguish between certain shades.

Furthermore, the eye is only sensitive to a short range within the electromagnetic spectrum, namely the visible region, between 400 and 700 nm.*

Barry McLow
3 Buxton Lane, Droysdon, Manchester M43 6HL
barrymclow@hotmail.com

I. Thorn and C.O. Au (eds.), *Applications of Wet-End Paper Chemistry*,
DOI 10.1007/978-1-4020-6038-0_10, © Springer Science+Business Media B.V. 2009

Table 10.1 Wavelength of light and the colour which absorbs it

Wavelength of light (nm)	Colour of this light	Colour which will absorb this wavelength
400–435	Violet	Greenish yellow
435–480	Blue	Yellow
480–490	Greenish blue	Orange
490–500	Bluish green	Red
500–560	Green	Purple
560–580	Greenish yellow	Violet
580–595	Yellow	Blue
595–605	Orange	Greenish blue
605–700	Red	Bluish green

It is important to realise that primary colours for light differ from those for colourants. Where light is concerned these three primary sources are red, green and blue (yellow light is not a primary light, being a mixture of red and green lights).

For colourants, the three primary sources are red, yellow and blue.

From childhood we are taught that white light is a mixture of coloured lights and this is proved to us all by the 'rainbow' effect or the splitting of white light through a prism.

Table 10.1 shows the different regions of the visible spectrum, the colour of light at that specific wavelength band and the shade of colourant which will absorb that particular wavelength.

Thus, for a yellow piece of paper, it is the wavelengths at the lower end of the spectrum, i.e. the violet and blue regions, which are being absorbed, and the wavelengths at the higher end of the spectrum, i.e. the red and green regions, which are being reflected. (Note that this example has been chosen to reinforce the fact that yellow light is a mixture of red and green lights as indicated above.)

For white and black papers, all wavelengths will be reflected, although not necessarily in equal proportions, the main difference being that with black papers, the degree of reflectance will be far lower compared to white.

Browns constitute a vast range of shades and as such it is not possible to make a definitive statement as to which wavelengths will be absorbed or reflected. This will be dictated by the shade itself.

All shades can be represented by a reflectance curve (Fig. 10.1), and this curve is very much like a fingerprint. When two curves are identical they will have the same shade, but, for example, not all reds will have the same-shaped curve.

10.2 Chemistry

10.2.1 Chemistry of Colours

The 'heart' of a dyestuff is the chromophore and consists of a large system of conjugated double bonds, alternating with single bonds, and this is known as the pi

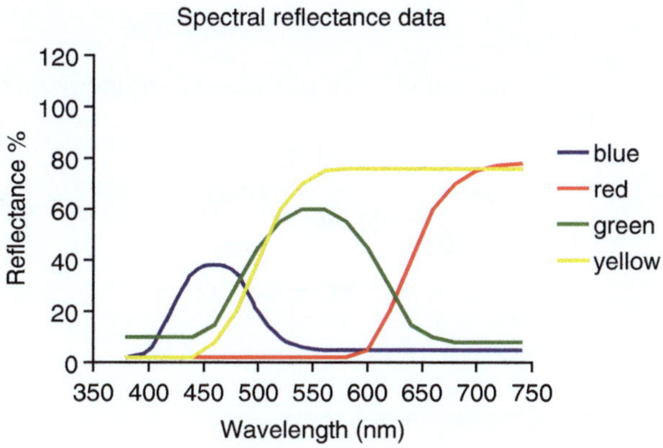

Fig. 10.1 Reflectance curves

electron system. In order to absorb light and qualify for the name 'dyestuff', a molecule should have a minimum of four conjugated double bonds.

When visible light encounters this pi electron system, the colour seen is a result of the degree of absorbance of the light by the pi electron system of the dyestuff.

As a structure, cellulose itself is planar and negatively charged and it follows, therefore, that there will be some repulsive forces between the cellulose fibre and any colourant which is anionic in nature.

However, the characteristically large pi electron system of anionic direct dyes provides a strong attraction between dye and fibre, known as Yoshida or van der Waals forces.

Whether a dyestuff is able to fix to the cellulose is dependant upon whether the ambient conditions allow for the repulsive forces to be overcome.

10.2.2 Types of Colourants Used Within the Paper Industry

The most widely used colourants for paper are:

Small, often non-planar anionic dyes, called 'acid' dyes
Small, often non-planar cationic dyes, called 'basic' dyes
Large planar anionic direct dyes
Large planar cationic direct dyes
Very large non-planar cationic polymer dyes
Organic or inorganic water-insoluble pigments

Acid dyes are generally small molecules, with a molecular weight of 350–500, and a small conjugated double bond system (Fig. 10.2), with no attractive forces. As such they have very little affinity for cellulose fibres. They do, however, have extremely high solubility and normally produce very bright shades. They can be applied at the surface and they are generally restricted to papers which have a limited use, i.e. crepe paper.

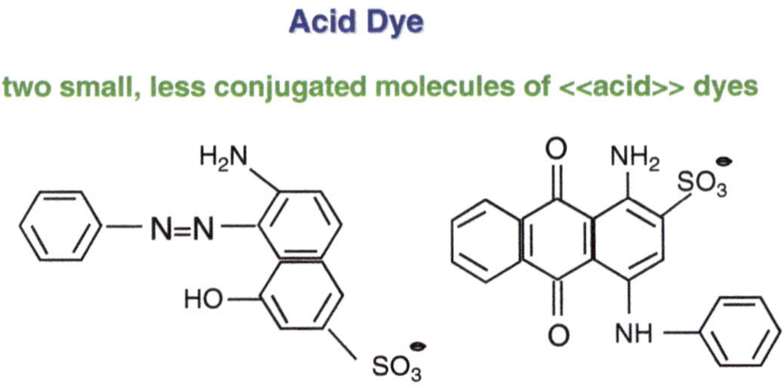

Fig. 10.2 Chemistry of an acid dye

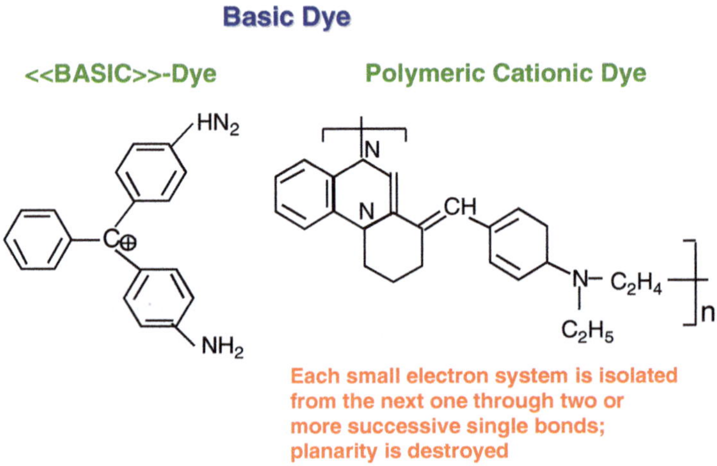

Fig. 10.3 Chemistry of a basic dye

Basic dyes, with a molecular weight of 300–500, are cationic in nature (Fig. 10.3) and generally speaking have low affinity for bleached fibres. They do, however, react with acid groups of lignin, whereupon fixation is both rapid and complete. While they produce brilliant shades and their dyeing costs are invariably lower than with dyes from other classes, their fastness properties are low. They are not water-soluble and require, in some cases, quite large additions of acid in order to produce a stable liquid dyestuff. Their use within the paper industry is declining, and is limited to papers containing a degree of mechanical fibres, such as newsprint, yellow pages and low-quality industrial wipes. Products typical of this group include dyes such as Auramine, Chrysoidine, Rhodamine, Methyl Violet and Victoria Blue, although they may also be known by a proprietary trade name.

Anionic direct dyes are the most widely used colourant within the tissue, printing and writing paper sectors. Having a typical molecular weight of 550–2,000, they produce a wide range of shades, although certain bright shades cannot be achieved using direct dyestuffs. Due to the diverse nature of the products which fall into this category, it is not possible to make definitive statements as to the properties of such dyes. They vary from products which have high light fastness to products with low light fastness, from products with high affinity to products with virtually no affinity. Typically anionic direct dyes will be planar (Fig. 10.4) in nature (exceptions being those dyes based on the phthalocyanine molecule). Factors which can have an effect on dyeing performance of direct dyes are dealt with in a separate section.

Cationic direct dyes are similar in shape to anionic direct dyes (Fig. 10.5), but while anionic dyes contain negative groups which have repulsive forces and impart water solubility to the product, cationic direct dyes have positive groups attached which allow an attraction between fibres and dyestuff. Accordingly, therefore, dye fixation with these dyes is excellent. This property does, however, lead to certain problems when these dyes are used either in conjunction with anionic dyes, or with furnishes which contain a degree of mechanical fibres or anionic trash (such as CTMP pulps).

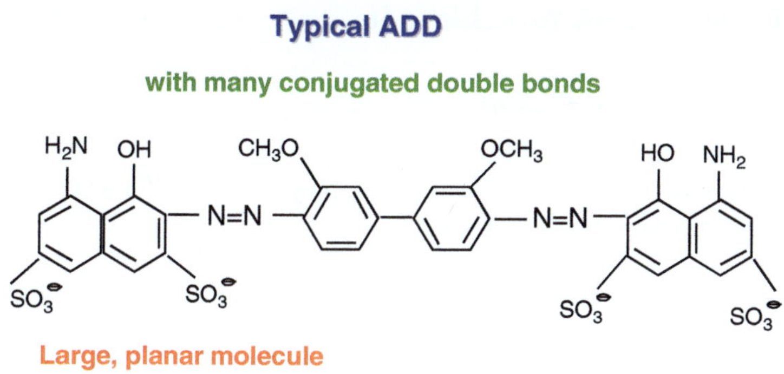

Fig. 10.4 Chemistry of an anionic direct dye

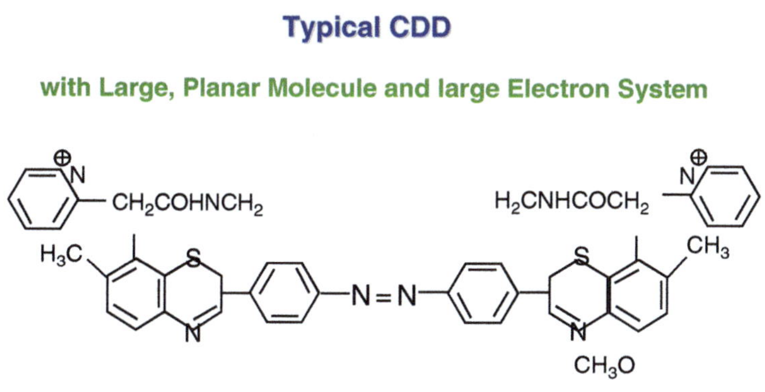

Fig. 10.5 Chemistry of cationic direct dyes

Generally speaking, the dyes have to be dissolved in acid to produce a stable dye solution.

Cationic polymer dyes are high molecular weight products (>2,000) whose affinity for bleached cellulose fibres is low. They will dye groundwood and mechanical containing furnishes and have slightly improved light fastness when compared to the conventional basic dye types.

Water-insoluble pigments have a somewhat limited, but nonetheless essential, part to play in the colouration of certain papers. They are normally applied as anionic dispersions. They have virtually no affinity for cellulosic fibres and instead rely upon either mechanical retention from the paper machine itself or flocculation via a cationic product (fixing agent) to produce small flocs which can be retained on the fibre surface.

Pigments have much brighter shades and higher fastness properties when compared to direct dyes; however, dyeing costs tend to be higher.

10.3 Colour – How to Achieve It

10.3.1 Selection of Colourants

Many factors have to be considered when choosing a colourant for a particular paper, most importantly the end use of the paper.

Are there any specific fastness properties required of the paper?

Colour fastness of paper is the ability of the paper to withstand certain conditions and reagents it is likely to come into contact with, such as water or adhesives or dry cleaning fluids. An important property is fastness to light, where certain papers, such as book covers, are required to retain their shade even after prolonged light exposure. By contrast, toilet and facial tissues require no such properties and consequently can be produced using dyes which are adversely affected by light.

Of increasing importance are papers likely to come in contact with food. In such papers could any dyestuff be transferred to the foodstuff itself? Is the paper a 'use once and then dispose of' type of paper, e.g. a tissue? Does the paper need to be even-sided in terms of the shade of each side of the paper?

Typically, the types of paper indicated in Table 10.2 would require the following properties, and be coloured using the type of dye indicated.

One problem which can arise with colour is a need for the paper standard (reference) and the production quality to be non-metameric. Metamerism is a phenomenon which exists when two samples of paper which appear similar under one light source appear to be different when viewed under a different light source. The only way in which this can be overcome is to use the same dyes in the production that were used in the standard itself; however, by careful selection of colourant (and this requires the use of a computer-based program), this phenomenon can be either reduced or overcome.

Table 10.2 Type of dyes used for different paper products

Paper type	Properties required	Typical colourant
Industrial wipes	None	Basic dyes or direct dyes
Toilet and facial tissues	Bleachable	Direct dyes
Printing and writing	None	Direct dyes
Cover papers	Good light fastness	Some direct dyes, plus pigment pastes
Laminate base papers	Excellent light fastness	Specific pigment pastes
Laundry tag papers	Fastness to dry cleaning process	Specific pigment pastes

10.3.2 Application of Colour

There are several means by which colour can be applied to paper but the main methods are:

Mass application, either by batch addition, continuous addition, or a mixture of the two; or Surface application via either the size press or coating.

The majority of colour is applied at the wet-end of the machine.

Papermaking is a continuous process and the ideal scenario is to apply the colour in a similar, continuous manner. Continuous addition can be achieved by metering colours into the system, as late as possible, at such a point that allows both mixing and fixation of the dye to take place, but avoids colour being added which would result in colouration of a holding chest. A typical addition point would be the exit of the machine chest prior to dilution of the stock at the fan pump. By adopting such a system it is possible to make rapid changes to the shade of the paper being produced.

These shade changes can be made in one of two ways, by either manual or automatic adjustment of the dye metering pumps. The first method relies upon the machine operator being able to accurately assess the change in shade required and adjusting the dye addition accordingly, while in the second method, the shade of the paper being produced is continuously instrumentally measured against the required standard. Within the online colour control system, the 'build-up' characteristics of the colourants are recorded, and the additions are automatically made so the shade is adjusted from that obtained to that required.

It is important to know that this latter system cannot be used until the shade of the paper being produced is relatively close to that required.

When both anionic and cationic dyes are being used to produce a particular shade, these colours cannot be added at the same point or via the same dye addition due to co-precipitation of the dyes. In such circumstances the anionic dyes, which have lower substantivity, would be added first to the thick stock, whilst the cationic dyestuffs, due to their high substantivity, can be added later, even after dilution to the lower consistency required for sheet preparation.

Continuous dyeing has its limitations in that certain deep colours or fastness properties cannot be achieved due to the restricted dyeing time involved.

Batch addition uses longer contact time which allows for far more complete fixation of the dye, coupled with the effect of higher temperature which will enhance the performance of the dye.

Furthermore, powder dyestuffs which could only be used in a continuous system after prior dissolution can be added dry as sufficient time exists for the powder to dissolve. Batch addition of colour has the disadvantage that any errors in colour addition cannot easily be corrected and the possibility of off-shade paper being produced is high.

For deep shades, therefore, a compromise can be achieved by using a mixture of the two types of colour addition, namely split addition. Split addition requires approximately 80% of the colour being added batch-wise, with the remaining 20% being added continuously. Such a method allows for correction of off-shade paper to be made. However, it must be remembered that as the majority of dye is being added outside the control of the continuous dye metering system, any automatic colour control via this point of addition needs to be closely monitored.

Surface colouration via the size press can be achieved. However, the fastness properties of the finished paper are likely to be low. Shade differences can be quickly corrected as the time between dye addition and reeling is very short.

Acid dyes which have no substantivity, but very high solubility, can be used in this way.

10.4 Factors Affecting Colourant Choice and Performance

Pulp freeness

Increases in refining will increase the depth of colour produced.

Furnish variations

Varying pulps will produce differing degrees of colour strength. Potentially, the lowest dyeing performance will be seen when mechanical pulp is incorporated into the furnish. This is caused by the anionic nature of the pulp increasing the repulsive forces which exist between fibre and dyestuff. This can be overcome by the prior addition of a cationic fixing agent.

Consistency

Colourants should be added at the maximum consistency that allows for thorough mixing of the furnish and colourant to take place. Preferential dyeing of fibres can take place if dye and a limited portion of the fibre are kept in contact for any period of time. By adding colourants, particularly anionic direct dyes and pigments, at a lower consistency, the Yoshida and van der Waals forces are less effective and dyeing potential is lower. Cationic direct dyes can be added to thin stock as an attraction between dye and fibre exists.

Water hardness

Increases in water hardness will help to reduce the repulsive forces which exist between fibre and dyestuff, and help to improve dyestuff retention. Mills whose water supply is soft will often artificially increase water hardness by addition of such products as magnesium sulphate.

Filler

Additions of chalk or china clay will increase the amount of colourant required to produce a particular shade. How much filler can be added will depend upon the physical characteristics required of the paper, and the likely increase in cost. Certain deep shades cannot be economically produced if filler content exceeds 10%.

Broke

Broke has to be carefully controlled to ensure any such addition does not make the required shade unattainable. When using wet, bleached broke which has been treated with sodium hypochlorite, it is essential to ensure the bleach is 'antichlored' by the addition of a suitable chemical.

Fixing agents

Fixing agents are cationic products which are added to give complete fixation of the colourant when ambient conditions prevent this occurring naturally.

As these products vary in chemistry, their point of addition in relation to the colourant, i.e. pre- or post colour addition, is very much product-specific.

When used as flocculating agents for the retention of pigment pastes, the fixing agent is added prior to the pigment. Cationic fixing agents can also be used to neutralise anionicity inherent in mechanical pulps. In such instances the fixing agent should be added prior to the addition of colour. It should be noted however, that fixing agents can have the effect of producing both a duller shade and a drop in the light fastness of a particular paper.

Sheet moisture content

Certain products, particularly anionic direct violet dyes, are affected by the moisture content of the dried sheet. Paper containing these products will change shade as moisture content increases. When using these colours it is important to ensure that the paper is fully conditioned before assessment against the standard quality is made.

10.5 Colour Measurement – the CIE L*a*b*C*h System

Figure 10.6 indicates how the colour of a specific sample of paper can be represented.

Every colour has its own position within the model and its position relative to that of the required standard can be determined.

The model consists of three axes:

L*, the lightness/darkness of a sample of paper
a*, the degree of redness/greenness of the paper
b* the degree of yellowness/blueness of the paper

For this reason the a* and b* axes can have both positive and negative values. Using this model we can see that neutral shades such as whites, greys and blacks will be positioned around the L* axis. It should be noted that not all white papers will have an L* value of 100; nor all blacks an L* value of 0. Once a specific colour, such as a red, a yellow, a blue, etc. is added, two things are likely to happen:

The L* value will reduce as the amount of light reflected from the sample is lower, and the a* and b* co-ordinates will change in accordance with the colour added.

Simplistically, if a red colour is added, this will result in an increase in the +a* value, whilst a blue colour will result in an increase in the –b* value.

C* is a measure of the Chroma or brilliance of a paper.

The higher the chroma, the brighter the shade and the further away from the L* co-ordinate will be its position in the chart.

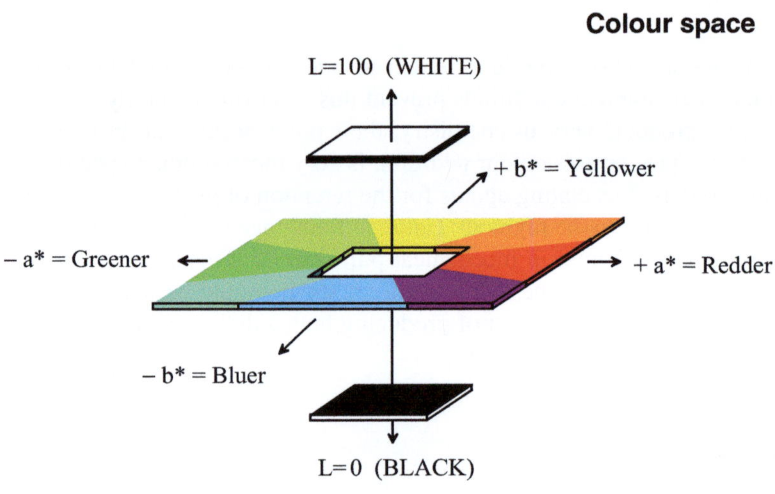

Fig. 10.6 Representation of colour

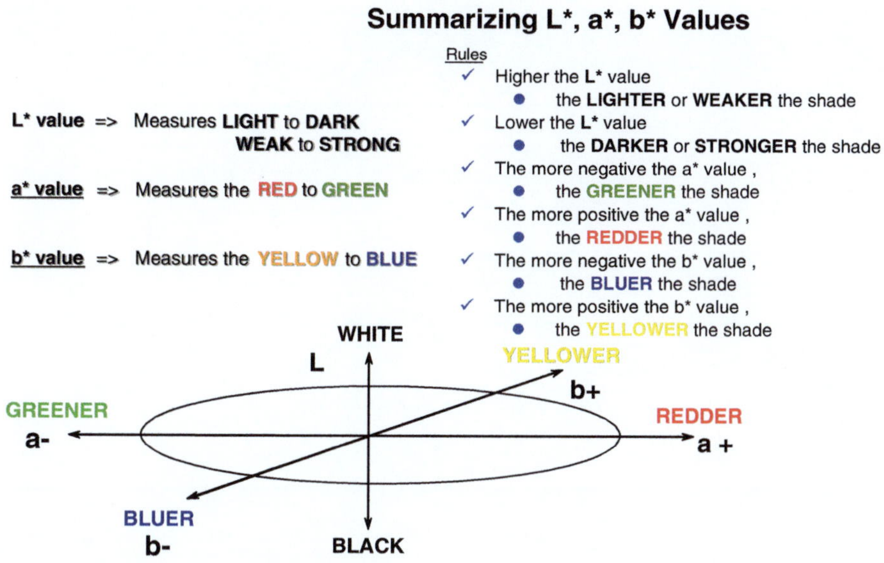

Fig. 10.7 A summary of the model

h is the hue angle of the paper. Although the model is irregular in shape, we can specify shades as though the model was circular, with 0° beginning at the +a axis continuing round until 360° ends up at this same point; thus a paper with a hue angle of 90° will be yellow, and a paper with 270° will be blue.

Hue angle offers a simple means of determining the difference in shade of two papers. If, for example we take two orange papers, the one which has the higher hue angle will be yellower in shade. The model is summarised in Fig. 10.7.

In producing coloured paper, manufacturers will have specific tolerances which are permissible, and providing the co-ordinates of the paper being prepared are within those limits, the paper would be deemed to be satisfactory. When paper is produced which is outside the tolerance limits set by the end user, the paper will be reused within the paper mill as 'broke'.

Broke has to be carefully used to ensure that it does not impact on any shade, thus making it impossible to obtain. For example, pale-coloured broke can be used in a paper of similar hue, or even as a constituent in another shade providing the colour it contributes is beneficial to that shade. A pink-coloured broke can easily be used in a red paper, or a pale yellow broke in a green paper.

Where problems will exist, and lead to the mill producing even more broke, would be, for example, when red broke is included in a green shade. Red is a complementary, or opposite, colour to green. The inclusion of broke here would make the green shade duller, and in some instances, impossible to obtain.

The amount of broke which could be used is limited to that which would still allow for three colours to be used to obtain the required shade.

Chapter 11
Online Monitoring and Control in the Wet-End

Roland Berger

11.1 Introduction

From its very beginning, papermaking has been targeted to produce a homogeneous sheet structure out of large volumes of water containing small amounts of fibre and filler. Fibre concentrations of below 1% are quite common in this context.

In order to successfully control both the desired paper properties and the production process, papermakers introduce functional additives and process chemicals into the fibrous suspension. However, these chemicals have to interact with the fibres if positive results are to be obtained.

What impedes interaction are the extremely unfavourable conditions prevailing in fibrous suspensions, such as very high dilution levels, strong turbulence and high shear forces, short contact times with fibres and, not to forget, anionic trash – all of which counteract optimal adsorption of chemicals onto the fibres.

Papermakers have long been aware of the fact that a number of measurable parameters (cationic demand, Zeta Potential, COD, turbidity, conductivity, pH) enable characterisation of the pulp suspension in terms of its colloid chemistry. However, this takes a lot of time and effort. In order to obtain representative results these often dynamic and complex processes should best be monitored continuously. This is where dedicated measuring devices are needed which enable paper manufacturers to respond in time to changing system conditions.

R. Berger
Product Manager Mütek Lab & Online, BTG Instruments GmbH, Arzbergerstrasse 10, D 82211 Herrsching, Germany
e-mail: roland.berger@btg.com

I. Thorn and C.O. Au (eds.), *Applications of Wet-End Paper Chemistry*,
DOI 10.1007/978-1-4020-6038-0_11, © Springer Science+Business Media B.V. 2009

11.2 Elements of a Pulp Suspension and Various Measuring Parameters

To investigate and understand chemical reactions in the papermaking process a number of measurable parameters are used. Figure 11.1 shows an overview of the various fractions in a pulp suspension and the suitable measurement technique. The pH value represents H^+ and OH^- concentrations, conductivity represents the amount of dissolved ions in solution, the charge of colloidal dissolved and fine dispersed material is detected with cationic demand, turbidity in filtrate focuses on fine dispersed material and the Zeta Potential characterises the surface of fines, fillers and fibres. Colloidal chemistry takes place within these fractions and can be understood by using different lab measurement techniques. The parameters pH, conductivity and turbidity are widely known and will not be discussed further. With cationic demand and Zeta Potential measurement we have the possibility of analysing two types of charges leading to a new dimension of characterising the pulp suspension.

Like the majority of fillers, paper fibres by nature carry small negative charges. Utilising this fact modern paper chemistry causes positively charged particles to adsorb onto the fibres via polar bonds. This is why practically all paper additives contain positively charged particles.

Polar bonds offer the dual benefit of being strong and they build very quickly. Measuring the charges of both chemical additives and fibres is crucially important in order to evaluate, control and exploit polar bonding for the purposes of paper

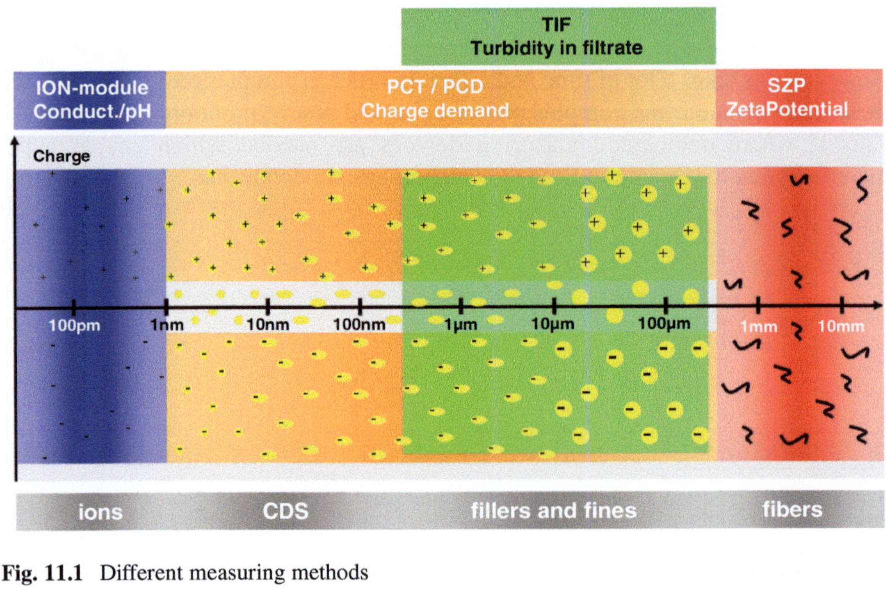

Fig. 11.1 Different measuring methods

manufacturing. Thus measuring charge using a Mütek™ PCD-03 Particle Charge Detector and a Mütek™ SZP-06 System Zeta Potential provides a clear picture.

11.3 Charge Measuring Principles Based on Streaming Potential

Even some 20 years ago, charge measurement was not common in the paper industry. At that time, Zeta Potential had long been known to be measurable but its nature and benefits were still not widely spread among users.

11.3.1 Zeta Potential

A particle, such as a paper fibre, in an aqueous system may exhibit positive or negative charge carriers on its surface. These charges cannot, however, be measured as a whole, because the particle is surrounded by a diffuse layer of counter-ions that weakens the particle charge.

This diffuse layer, the so-called double layer, consists of a number of ions and a number of ions of opposite charge which are in part rigidly held on the particle and in part more loosely bound further away from the particle. The double-layer model around particles is taught at papermaker's schools and universities and is well known.

If the loose or diffuse double layer is removed, for example, by shearing, charge measurement becomes possible. However, the charge is slightly reduced by the dense layer surrounding the particle. This weakened and measurable particle charge is termed Zeta Potential.

Basically, it is possible to determine the Zeta Potential of both dissolved and dispersed particles. In a mill environment, however, Zeta Potential measurement of dispersed particles such as fibres and fillers is most widely used today.

11.3.2 Cationic Demand

Similar to solid particles, colloidal dissolved substances, too, build up an ion cloud corresponding to the double layer model. In the case of colloidal dissolved substances, it is extremely difficult to shift the ion cloud in a reproducible manner for subsequent charge determination. This is why another method, the so-called charge titration, has been established in the paper industry.

During charge titration, a polyelectrolyte is added to the solution and forms a strong bond (simplex) with the oppositely charged particle. In this manner, the particle charge is continuously reduced to the point of zero charge.

The charge of the dissolved particle is then derived from the amount of consumed polyelectrolyte. Since this method is most frequently applied to identify the level of dissolved anionic trash, it is often termed 'cationic demand measurement'.

11.4 Approach to Control Chemicals Based on a Wet-End Charge Survey

Charge measurements give meaningful information about surface interactions, attraction forces and adsorption mechanisms in colloidal systems. In a mill environment, lab charge measurements primarily meet the following objectives:

- Identification of critical sources of interfering substances in a stock system
- Establishment of a charge balance of the paper machine
- Assessment of the stability of colloidal systems
- Quantification of specific disturbing substances
- Selection and performance assessment of various fixatives

11.4.1 Wet-End Charge Surveys at a Paper Machine

In order to understand the reaction of each additive dosed into the process, one should first assess the overall charge conditions prevailing in the paper machine wet-end. This is done by conducting a complete, so-called wet-end charge survey. The Particle Charge Detector measures the charge of dissolved matter in the water phase, whereas the System Zeta Potential measures the actual charge of the fibre surface. It is important to assess the charge characteristics of all incoming furnish streams, after mixing and dilution points as well as before and after chemical additions. This allows evaluating the efficiency of the chemical dosed. Either the additive is attracted onto the fibres and fibre fines creating the desired paper property or it stays in the water. Figure 11.2 shows a schematic of a wet-end charge survey.

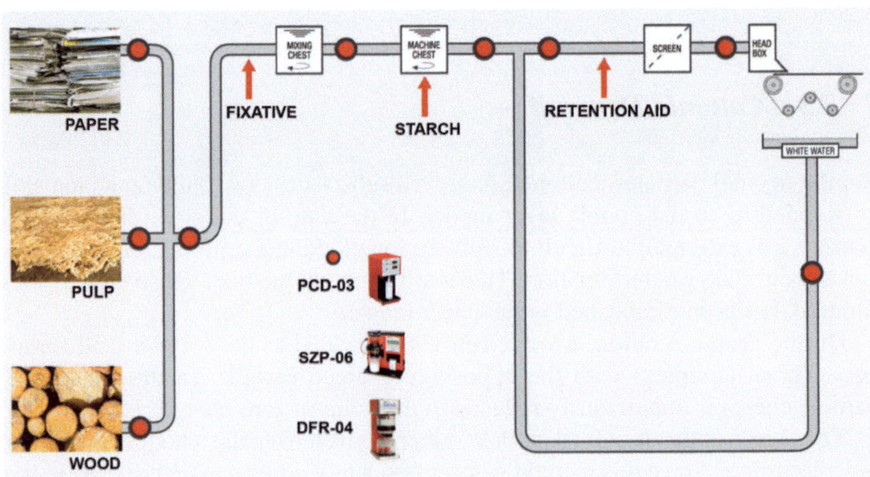

Fig. 11.2 Practical process survey with lab instruments

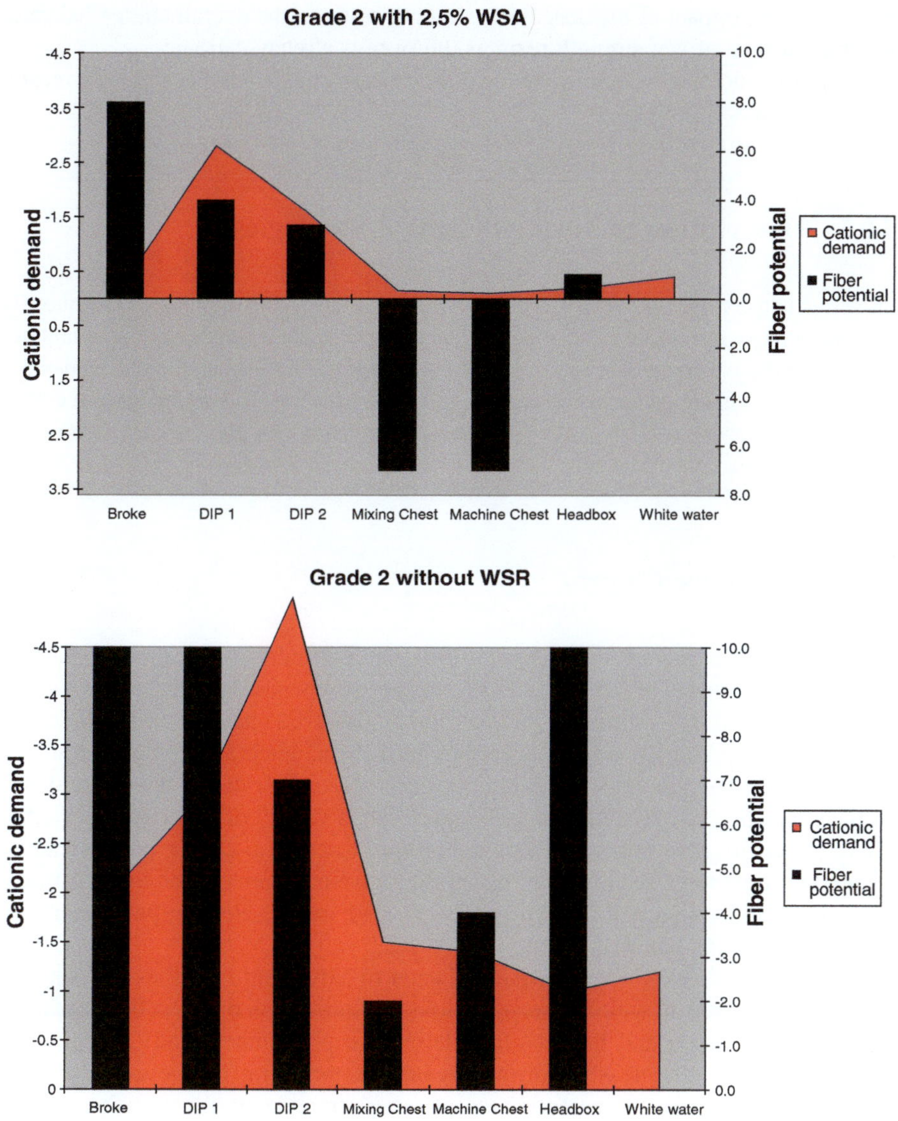

Fig. 11.3 Example of a wet-end charge survey at a tissue machine

Figure 11.3 shows two examples of a wet-end charge survey. The red area illustrates the variation of the anionic trash level throughout the process. The black columns show the development of the absorption capacity of the fibres from raw material lines to the white water.

This example indicates several wet-end chemistry aspects:

- The final paper wet strength strongly depends on the variation of the incoming furnish

- The strong impact of the dosed wet strength resin on the overall charge balance
- Over-dosage of wet strength resin as the fibre is already cationic
- A cationic demand very close to the point of zero charge which can be the reason for production problems

11.5 Description of Various Control Strategies

Dynamic and complex processes as pulp suspensions have to be monitored continuously if representative results are to be obtained due to paper grade changes, varying quality of raw materials, and chemical carry over from bleach plants. Dedicated measuring devices are needed to enable paper manufacturers to quickly respond to varying system conditions. The following section explains various control strategies and their applications in different grades.

11.5.1 Fixative Control

In the wake of an increasing closure of mill water circuits, mill technicians notice a loss of performance of paper additives and process chemicals. This negative effect is mainly caused by a variety of interactions of dissolved chemicals with dissolved anionic trash primarily originating from broke and integrated pulp operations, which accumulated in the white water. For this reason, various fixatives are dosed in the paper process. Chemical suppliers provide a variety of fixatives adjusted in charge intensity and molecular weight. Before closed-loop charge control can be applied it needs to be secured that the right fixative has been chosen using cationic demand measurement and deposition testers. The major selection parameters are cost efficiency and deposit tendency.

Basically, we have to differentiate between two different functions of fixatives: (1) a cationic fixative, which adsorbs colloidal anionic trash onto the fibres so as to avoid agglomeration and the resulting deposition phenomena; and (2) a fixative with high cationic charge, which drastically impacts the overall charges of the process that greatly influence the performance of other strategic additives such as sizes, starches and retention aids.

This is where online charge control becomes indispensable. When an appropriate fixative has been selected, charge control effectively stabilises the charge household of a system. For this purpose, a positively charged polymer is added to neutralise negatively charged interfering substances. In case the raw material carries high trash loads, fixative dosages have to be increased, whereas they must be reduced for low-load raw material of a superior quality. This strategy avoids over-dosage of fixatives which would diminish the fibre charges, thus impairing the adsorption of cationic additives at the fibres. By preventing over-dosages, papermakers may reap cost benefits and enhance the effectiveness of subsequently added cationic aids. As charge levels are stabilised, subsequent chemicals will perform more efficiently, thus contributing towards an optimised paper quality. This is true for all kinds of additives, irrespective of whether they carry high charges (e.g. highly cationic starches) or low charges (e.g. faintly cationic polyacrylamides). Without charge stabilisation, a highly charged additive may react directly with strong anionic trash in the process, thus giving rise to deposition phenomena and inconsistent paper qualities, such as varying dry strength properties.

Similarly, a polyacrylamide with low charge used as a retention aid will show a non-uniform performance in response to varying levels of anionic trash. The reason for this is that charge variations affect flocculation processes. Both the bonding strength and the reaction rate of additives and fibres are strongly influenced by charge variation.

The Mütek™ PCT-20 Charge Analyzer together with the Mütek™ TSS-71 Filtration Sampler measures the charge demand at critical points of the wet-end. It is important to measure the cationic demand of the filtrate after the dosage point of the fixing agent. This allows for a tight closed-loop control based on the colloidal anionic material at these locations. Direct measurements in white water or flotation waters can be used for monitoring the process, e.g. avoiding over-dosage of cationic additives like cationic starch or wet strength resin. In very few cases, the cationic demand of the unfiltered white water is used to control cationic fixing agents in the thick stock area.

11.5.2 Retention Aid Control

Retention aid control is based on two or more measurement points around the headbox. When measuring total consistency and ash consistency, a balance for retained material can be made. Classically, the retention aid is dosed according to the total consistency in the white water as white water is used in several locations of the paper machine for dilution. Overall benefits of this type of control are a more stable basis weight, a more stable moisture content and more stable caliper. This strategy has been proven beneficial in fast running machines with a limited amount of grades. The drawback of this control strategy is that variation in retention aid dosage has an impact on formation and drainage. Additionally, it cannot compensate ash variability.

Before closed-loop control with a Mütek™ RET-20 retention control can be started the best suitable additive combination has to be selected. For this reason numerous measurements for drainage and retention should be performed. Historically this has been done by Britt Jar measurement and modified Schopper Riegler. Automated laboratory units like the Mütek™ DFR-04 Drainage Freeness Retention can nowadays be applied for faster selection of the retention aid performing in the best way.

11.5.3 Ash Control

Stable ash content in the sheet usually is controlled with the dry ash scanner value dosing filler in the thin stock. Often this control strategy is limited when coated broke is used in varying amounts. Another source of disturbance can be the retention aid control loop. A retention aid controller reacts with increased polymer dosage by, for example, variation in anionic trash loads. This will quickly impact the ash control loop and a stable sheet ash cannot be maintained.

Thick stock ash control is another promising solution. An ash sensor either based on dilution of the blend chest stock or an inline sensor in various thick stock locations – coated broke, uncoated broke and blend chest pump enable control of thick stock filler addition or ratio control of incoming pulp streams.

11.5.4 Gas/Air Control

Most paper machines have measuring equipment installed to monitor free air in the white water or headbox. To successfully control the defoamer/deaerator dosage, free and dissolved gas need to measured. The Mütek™ GAS-60 Gas Analyzer provides two measurement points as recommended for a stable control loop. The chemical deaerator is dosed based on the free or dissolved gas content in the white water. The free or dissolved gas content of the headbox allows supervision of the mechanical deaerator, monitors leaking pumps and pinpoints other disturbing sources of air.

11.5.5 Combined Control Strategies

Independent control loops, e.g. thick stock charge control and retention aid control, can improve the overall stability of the paper machine. Figure 11.4 illustrates how fixative control based on the filtrate charge of the thick stock can improve retention aid control. Part 1, without charge control, shows strong variability in retention aid dosage while maintaining stable white-water consistency. Part 2, with implemented charge control, shows that variability in retention aid dosage is reduced. Consequently with charge control retention aid control can now be applied over a much longer period of time – even during grade changes.

11.5.6 Multivariable Control (MVC)

The strategy most widely used in the wet-end today is single-loop control. One input variable, e.g. total consistency controls one output variable, e.g. retention aid dosage. Due to the fact that most parameters are impacted through various variables,

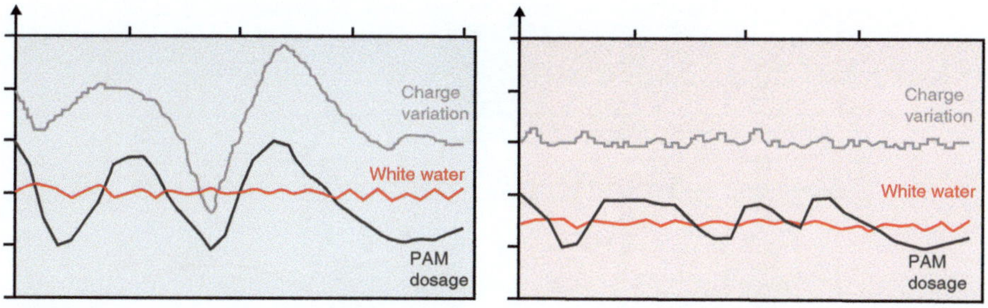

Fig. 11.4 Combined control strategies: Part 1 – without charge control (*left*), Part 2 – with charge control (*right*)

so-called multivariable control gains greater awareness. Multivariable control needs to be implemented by process control experts and is labour-intensive. MVC needs to be adjusted over time due to changes in the process, e.g. different chemistry which makes MVC maintenance-intensive. Additionally, this type of controller needs to be optimised and supervised using appropriate measuring technology.

11.6 Applications in Different Paper Grades

11.6.1 Fine Paper

Fine paper mills historically were the pioneers in implementing process control equipment in the wet-end. The reasons are various grades, coated broke usage and a wide range of chemical additives applied. Continuous charge measurements enabling controlled fixative additions to coated broke and blend chest substantially stabilise the charge household of the stock system. At the same time, the PM runnability (deposits, holes, web breaks) is improved. As described in Section 11.5 direct correlations between thick stock filtrate charge and the retention on the paper machines can be often found. Thick stock ash control allows for ratio control of broke lines and the possibility to maintain the desired ash level in the sheet. With defoamer control negative impacts on sizing are avoided and cost savings are achieved. Total consistency measurement points around the disc filter makes troubleshooting and higher efficiency of the separation process possible. Figure 11.5 shows a layout of instrumentation typically applied in coated fine paper production. A similar layout is applied for surface sized grades and uncoated grades when higher variability of pH, conductivity and colloidal dissolved material is derived from integrated processes.

Overall, the benefits in coated fine paper may be summarised as follows:

- Continuous monitoring enables a better understanding of process mechanisms.
- Over-dosages are avoided.
- Fixative is added on demand with cost-cutting effects.

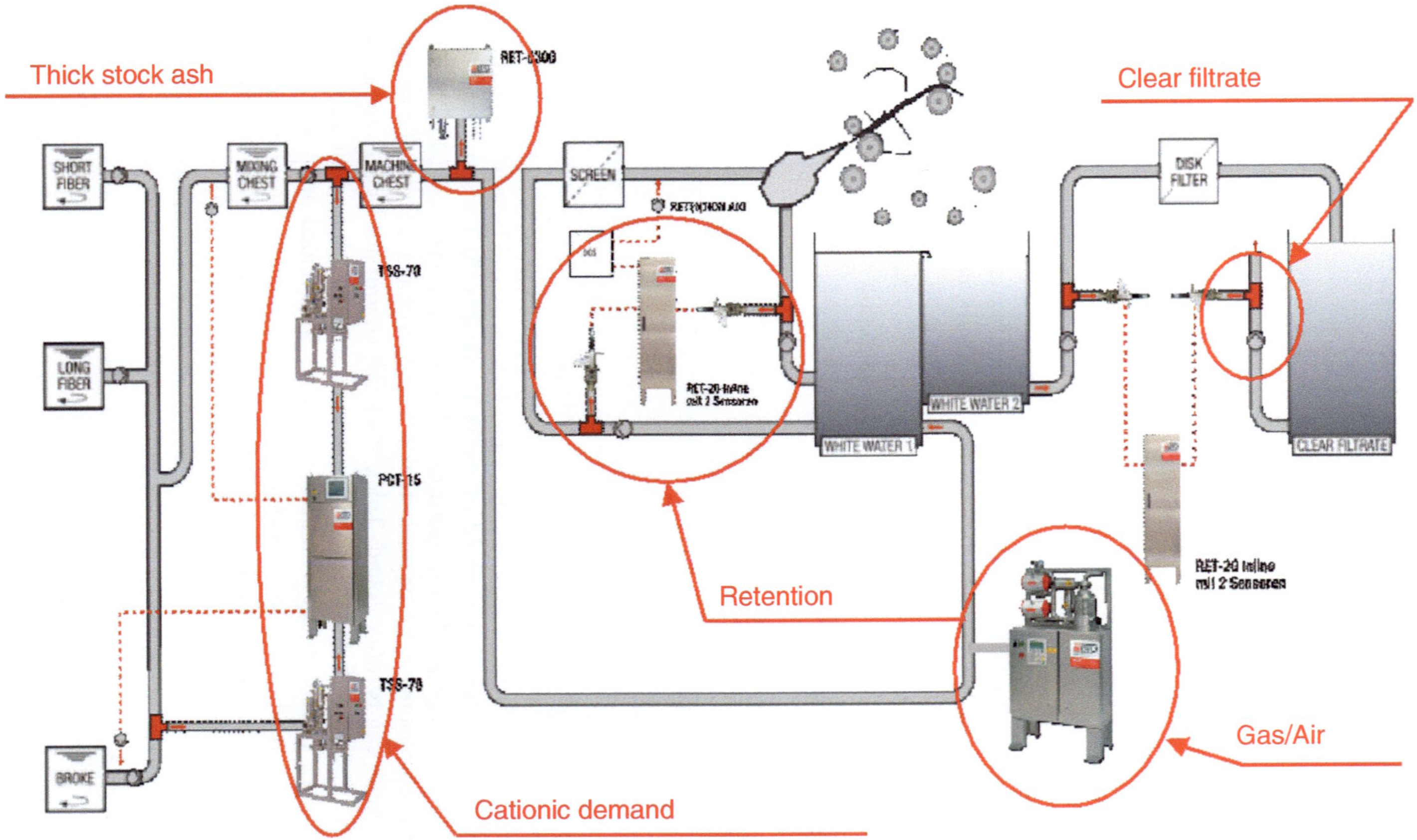

Fig. 11.5 Control approaches widely applied in fine paper mills

- The charge household is balanced and strategic additives are optimised.
- Deposits in offset press roll are avoided.
- Deposits in the pre-dryer section and the cylinder pick are more effectively controlled.

11.6.2 Newsprint

Today, the use of cationic fixatives in the production of recycled-fibre-based newsprint is state of the art in modern paper mills. Charge control favourably influences the performance of other chemicals, too, because this strategy excludes DIP variation. A typical observation in this context is improved filler retention, overall more stable retention aid control and reduced deposition propensity in the PM which contributes towards an optimised machine running.

Figure 11.6 shows an example of an optimised controller for fixative addition based on online cationic demand measurement. In wood-containing newsprint grades sometimes this control strategy is replaced by turbidity control of the filtrate of the TMP or blend chest. Due to the nature of turbidity measurement – dependency on filtrate quality – this control strategy is only applied when polymers with low charge density but high molecular weight are applied.

After optimising the controller, the colloidal charge of 1.50 ml after the machine chest is commercially and process-wise the most appropriate level (Fig. 11.6). Fixative additions of 150 g/t and 450 g/t are chosen as minimum and maximum dosages, respectively. If the measured cationic demand keeps below the target value for a longer period, the minimum dosage is added.

Besides charge control, retention aid control and defoamer control based on online measuring technique are state of the art in these fast-running machines. A small range of grades allows very smooth and stable control.

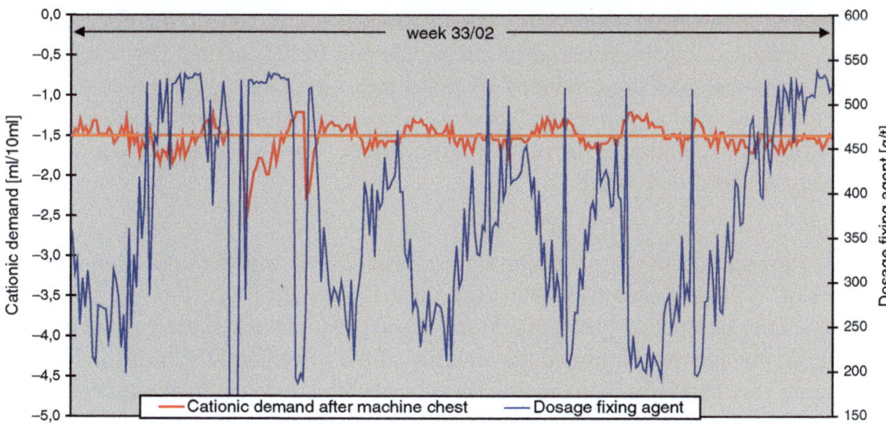

Fig. 11.6 Optimised controller for fixative addition based on cationic demand measurement

11.6.3 Specialty Paper and Tissue

In specialty paper and tissue production, online measuring technique has often not been used in closed-loop control in the past because of many grade changes and complex process chemistry. Small production volumes and slow running machines were not supporting acceptable payback calculations for investments in measuring technology.

This situation has changed. The use of a lab Zeta Potential as a fast and direct measurement opens new possibilities to investigate strategic additives and to better understand wet-end chemistry reactions. More cost-intensive additives are applied to meet the desired market demands allowing faster paybacks through process optimisation. Online measuring techniques are needed for automatic control of chemical additives and monitoring the wet-end dynamics. The following three examples demonstrate what can be achieved in tissue and specialty production.

Example 1
A mill producing kitchen towel and bathroom tissue has several issues with formation, deposits and a stable wet strength in the sheet. It uses 100% virgin pulp. Carboxymethylcellulose (CMC) is dosed to increase the anionic charge of the fibre to increase adsorption of cationic wet strength resin (WSR). This strategy is often found to be not effective. Additive costs are high, deposits can occur, and even if wet strength levels in the sheet can be increased process variability has not been diminished. One reason for this is that anionic CMC, dosed to the process also remains in the water phase. This can be seen by cationic demand measurement directly after the pulper after CMC dosage. After cationic WSR is dosed it will react with the anionic CMC. To get the desired wet strength, it is fundamental that these CMC–WSR complexes are retained in the sheet. If not, these complexes will result in deposits and increased COD and AOX values. Using retention aid to retain these complexes and to stabilise fines content in the white water should get more common. For best results closed-loop retention aid control will be needed.

In WSR-containing grades process charge sometimes can swing between anionic and cationic. A cationic process indicates free WSR in the water which is not retained in the sheet. Generally, when a process is too close to the point of zero charge, uncontrolled flocculation of chemicals and fibres occurs. This causes formation problems and the tendency to build deposits. Online charge control can provide the security to run a process close to neutral without the disadvantage of uncontrolled flocculation at the point of zero charge. This will lead to the maximum efficiency of the dosed WSR.

Example 2
Other examples are tissue products with a large DIP component. Often an excess amount of WSR is dosed to achieve the desired strength properties. Deposits and foam are daily production problems. Many tissue producers are starting to investigate cationic fixing agents to improve the stability of the incoming DIP furnish. This has to be done very carefully. Too much fixing agent will reduce the bonding sites on the fibre which are necessary for WSR adsorption. Only in times of higher anionic trash

loads from the DIP plant those additives should be used. To keep the optimum charge level in the DIP line the fixing agent has to be dosed according to a continuous cationic demand measurement of the filtrate in the thick stock. To optimize polymers in the flotation processes and to get the highest efficiency of wet strength resin defoamer chemicals should be controlled online on a lowest possible level.

Example 3
In mills which produce consumer tissue using mixed furnish and producing many different grades felt deposit problems are daily occurrence. To find out the origin of stickies mills usually invest a lot of manpower over time. Applying online charge control in white water provides ready access to information about the driving parameters of the process. Increased, up to date process knowledge results in much less deposits. Additionally, reduction of wet-end chemicals is most likely.

The chemistry of different grades is reflected by the varying charge level in the white water. This is driven by furnish ratios and the different amounts of WSR and dyes needed for the production of specific grades. A continuous cationic demand signal allows the operator to adjust the process quickly to the new process situation. A database established over time will help to provide a guideline for the operators for every single grade. Cationic demand variations within one grade reflect the quality variation of the incoming furnish. This allows the purely on demand usage of fixing agents. Label paper production with extreme variation in broke quality would be a typical example for this type of application.

11.6.4 Board

Due to the favourable market position in board production the interest in wet-end control equipment is rising. Historically, the focus was only on drainage improvement on the wire and in the press section, as the bottleneck is usually found in drying capacities. Today, with lower basis weight productions and increasing demand on board properties this focus is shifting. At lower basis weights retention control strategies are implemented and closed-loop control of retention aids are becoming state of the art for fast-running machines. Reduced fibre quality and increase of recycled pulp result in reduced sheet quality and an increased demand in strategic chemical additives. Therefore, classical charge control for fixing agents in the thick stock area becomes more common as it has been in use already for many years in fine paper and newsprint production.

Defoamer dosage is considered to be a cheap investment. For every machine there is an optimum air content level concerning energy savings through pumps, de-watering properties of the pulp and efficiency of chemical additives. After this optimum value has been detected it needs to be maintained. This is only possible by measuring the free and dissolved gas content in the white water and headbox. Closed-loop control of defoamer chemicals in board is very straightforward and relatively easy to implement.

Of course, multiple layer machines make wet-end control more complex. The need to focus on major layers to achieve the best possible payback is obvious. To find out the best online control strategy for a certain board machine all aspects of the process should be analysed.

With a Mütek™ DFR-04 lab unit the effect of various defoamers on drainage can easily be tested. For optimum performance of retention and drainage aids a Mütek™ RET-20 Retention Inline and a Mütek™ DRA-50 Drainage Analyzer are best installed directly in the headbox and white water.

11.7 Summary and Conclusions

In most paper manufacturing processes, wet-end chemicals are added in fixed dosages which are usually optimised for a single process condition, regardless of variations originating from changing qualities of the incoming raw materials and broke lines. However, these inconsistent qualities lead to varying requirements of wet-end chemicals. Process stability and paper quality will significantly be improved by closed-loop control of chemical addition.

References

Gliese, Thoralf, m-real Hallein, Austria; Online charge measurement as an instrument for white pitch control. Presented at 6th Mütek Customer Days, Herrsching, 1999

Gliese, Thoralf, m-real Hallein, Austria; Online Ladungsmessung als Instrument zur White-Pitch-Kontrolle. Wochenblatt für Papierfabrikation 2000 (4): pp. 202–205

Berger, Roland, BTG Instruments GmbH and Stephan Gruber, Papierfabrik Palm GmbH& Co and Online charge measurements enable process optimization and automatic control of fixative dosages. Presented at TAPPI Spring Technical Conference, Atlanta, 2003

Gruber, Stephan, Papierfabrik Palm GmbH& Co and Roland Berger, BTG Instruments GmbH; Prozessoptimierung und automatische Regelung von Fixiermittel durch online Ladungsmessung. Wochenblatt für Papierfabrikation 2003 (11/12): pp. 700–711

Berger, Roland, BTG Instruments GmbH Charge control for improved tissue production. Presented at Tissue World, Miami, 2004

Index

MIX
Papier aus verantwortungsvollen Quellen
Paper from responsible sources
FSC® C105338

If you have any concerns about our products,
you can contact us on
ProductSafety@springernature.com

In case Publisher is established outside the EU,
the EU authorized representative is:
Springer Nature Customer Service Center GmbH
Europaplatz 3, 69115 Heidelberg, Germany

Printed by Libri Plureos GmbH
in Hamburg, Germany